THE POLITICS OF GLOBAL CLIMATE CHANGE

THE POLITICS OF GLOBAL CLIMATE CHANGE

♦ ♦ ♦

Patrick M. Regan

Paradigm Publishers
Boulder ♦ London

Published in the United States by Paradigm Publishers,
5589 Arapahoe Avenue, Boulder, CO 80303 USA.

Paradigm Publishers is the trade name of Birkenkamp & Company, LLC,
Dean Birkenkamp, President and Publisher.

Library of Congress Cataloging-in-Publication Data

Regan, Patrick M., author.
The politics of global climate change / Patrick M. Regan.
pages cm
Includes bibliographical references and index.
ISBN 978-1-61205-789-7 (pbk. : alk. paper) — ISBN
978-1-61205-760-6 (consumer ebook)
1. Climatic changes—Political aspects—United States. 2. Greenhouse gas mitigation—International cooperation. 3. Conflict of interests. I. Title.
QC903.2.U6R44 2014
363.738'74—dc23

2014026407

Printed and bound in the United States of America on acid-free paper that meets the standards of the American National Standard for Permanence of Paper for Printed Library Materials.

Designed and typeset by Straight Creek Bookmakers.

19 18 17 16 15 1 2 3 4 5

CONTENTS

PREFACE

The idea of writing this book came while I was developing a course that would include a five-week section on climate change. In trying to develop the outlines for the course, I came across the Clean Energy and Security Act of 2009 and the failed Copenhagen summit later that year. It seemed odd to me that a House of Representatives with a 75-vote Democratic majority only barely passed an environmental bill supported by the newly elected president, and then the president, who went to Copenhagen with permission from the House to come home with a climate change treaty, returned without one. The failure at Copenhagen required an explanation. Most of those explanations I could find involved either descriptions of the institutional arrangements for negotiating and enforcing treaties or some description of the dynamics of the bargaining process that pitted China and other large polluters against the United States in the summit negotiations. The vote in the House, however, suggested that the explanation might be found closer to home, in U.S. policy and those who make it.

Given the campaign theme of Obama's 2008 election (Change), and the overwhelming Democratic majority elected to Congress, the close vote on HR 2454 was telling us that underlying interests of individual congressional representatives, rather than a global good, were driving their votes. I assumed that cutting CO_2 emissions to stabilize our climate was conventional wisdom, and because of this, I had expected an easy pass on HR 2454. The puzzle of HR 2454 along with the failed Copenhagen summit became the organizing theme of the climate section in my initial class. By the end of the class, I had a research project in the making, and to complete it, I had to learn about climate and atmospheric science, hydrology, ecology, and biology, and I had to think more closely about the relationship between

American politics and world politics. As a scholar of conflict, I had expected much of my attention to be devoted to links between climate change and conflict. That didn't happen.

Part of this project involved asking myself if we'd ever faced another problem comparable to climate change. Collective action problems are ubiquitous and quite varied, but I couldn't come up with one that posed the kind of challenge this one does. I was talking with a chemist friend about how the quantities of CO_2 are greater than those of the carbon product that was consumed, and how carbon consumption pervades everything we do. He reminded me that we don't create carbon or oxygen atoms; we just move them around. And when we combine them in certain ways, we create a molecule, CO_2, that has nearly four times the mass of a carbon atom alone. Because oxygen and carbon are so instrumental to our lives, ridding ourselves of the problem of excessively high concentrations of CO_2 will require changes in the way we undertake this mixing process. We can't avoid it, but we could do it more efficiently. I couldn't come up with another natural or anthropogenic process that could rival this one, either in the difficulty of resolving its consequences or in the normalness of the process of creating it.

Having thought through this issue from many angles, I came to the realization that the climate change problem that results from our overproduction of CO_2 molecules is primarily a political problem. Climate scientists can tell us about the effect of increasing concentrations of CO_2 in the atmosphere and the processes by which this takes place; hydrologists can tell us about the effect of drought or rainfall on soil moisture and crop yields; biologists can inform us about the deleterious consequences of changing migratory patterns in response to changing weather patterns. No one can tell us how to implement fixes, because those fixes require political decisions and commitment.

If politics are the key to solving the climate change problem, there had to be a way to think about how politics sits at the core of the collective action problem. We recognize that the individual is unlikely to take concerted action absent altruism. So the traditional collective action problem has to be approached differently; politics has to be front and center in our thinking about when and why we act (or don't). Moreover, it became clear that this isn't simply a problem for individuals or international treaties, but one that requires involvement at three levels. As individuals, we will act out of self-interest, altruism, or coercion. But our elected officials display self-interested behavior tied to electoral cycles, and talk of jobs plays better to constituencies than any altruistic call to save the planet. It is also clear that Congress has to affect our individual behaviors, either through coercion or enticements, and until it gets on board, each individual has little incentive to act in ways that reduce impact on the climate. Even the informed and committed will drive a less efficient car than is available or live in a bigger house than is necessary. Consumption patterns are so pervasive that just understanding the issue is not enough to substantially alter behaviors. And international treaties will be necessary to generate compliance and commitment at the national level. This is what the meetings in Paris, Durbin, Copenhagen, and Kyoto are all about.

My foray into the world of climate change politics made me look at my own small corner of the world from the perspective of waste. If we have to cut our consumption without changing our lifestyles—the easiest way to maximize on personal preference and climate stability—eliminating waste and increasing efficiency will be critical to the solution. Most of us today, I suspect, don't think about waste. Waste comes in the form of overconsumption—inefficiency or methods that privilege convenience over health or necessity—and from careless use of energy—lights left on, DVRs that can't shut down, and heat too high. There is also a considerable amount of waste of electricity in the transmission from power plants to houses. The government estimates that up to 40 percent of electricity generated for consumption is wasted in the transmission process (EIA 2014). Even though eliminating or reducing waste will not be enough to mitigate our impact on the climate, it will have to be a significant part of the easy steps along that route.

Unfortunately, in researching and writing this book, I became more pessimistic than optimistic. If we do nothing, the changes to our climate will be extreme, but as long as those with interests other than a stable climate make it harder to take resolute action, we cannot do enough. In the current political environment in the United States, private interests of individual representatives hold sway over the public good. Sensible climate policy will have to filter to the top of the domestic political agenda for us to make any useful headway. I also came to realize that the problem is so great in magnitude that individual behavior alone cannot suffice to address it. Add to this that people in a relatively small number of countries have to realize the magnitude of the problem and act accordingly, and it looks even more daunting.

For time, space, research assistance, and editorial skills, I thank the Joan Kroc Institute for International Peace Studies at the University of Notre Dame for providing me with the quiet solitude of a visiting fellowship and then providing a sabbatical leave once they hired me. Much of the research and writing was completed during these periods. Students at Binghamton University and Notre Dame allowed themselves, unwittingly, to be the initial targets for my ideas. Arizona State University hosted me for a talk at which I took away as much as I imparted, and the Midwest Political Science Association invited me to give one of their Empire Series Lectures at which some pointed questions forced me to think about how to conclude the book. Idean Saleyhan, Michael McDonald, Debra Javeline, and Todd Sandler provided early comments on the core ideas, and Meithili Mitchell and Chad Clay provided stellar research assistance along the way. I owe a special debt to Umi Deshpande for help in making sure that my ideas translated effectively onto the printed page. This took amazing persistence and great capabilities; I appreciate both.

1

♦ ♦ ♦

THE POLITICS OF CLIMATE CHANGE

The world presents constant threats to human existence: war within and between states, nuclear weapons, famine, disease. Given these challenges, managing global political, social, and ecological environments is a constant struggle. But these are not all equally consequential struggles. The tempo of these threats waxes and wanes as we become more politically astute or socially inept. Human viability itself, though, is rarely at stake. New viruses that threaten millions of people reach a level of epidemiological equilibrium before the whole population is at risk. Avian flu, the West Nile virus, AIDS, the plague, all confront the human ability to resist, evolutionary processes to adapt, and the immunity of enough people—in biology or behavior—to prevent catastrophic impact on the planet. Avian flu required the slaughter of millions of chickens, development of vaccines, and in most instances, simple vigilance. The West Nile virus was kept in check by, among other things, mosquito repellant. International negotiations and sanctions are used to manage nuclear threats such as Iran's. Although not trivial to either the world or individuals, these threats require neither the attention nor the commitment of the average individual. Compliance costs people little, and there is no immediate risk perceived to their ways of life. The solutions and, to some extent, even the threats are, for most people, background noise in their daily lives. Climate change, however, is very different.

The year 2014 was the hottest year on record, bar none. There have been warmer periods, but not at a time of human record keeping, let alone during the time of the modern human species. In the past 20 years, our language has reflected the reality of, and the concern about, the effects of climate change. Carbon footprints, hybrid cars, low-energy lightbulbs, and carbon sequestering are just a few of the terms that are now part of our everyday vocabulary. It is no longer a strange idea that airline passengers can reduce their own carbon footprint with a fare increase that goes toward planting trees in some ecological preserve. Empirical research has, for years, demonstrated the anthropogenic contributions to the processes that generate climatic changes, as well as to the encroaching threshold beyond which these processes may be irreversible (Hansen et al., 1981). In the span of a single lifetime, we have witnessed the nearly complete melting of the glacial cap on Mount Kilimanjaro, the retreat of some of the world's largest glaciers, catastrophic weather events, and the long-considered drowning of New Orleans. Even though the flooding of New Orleans is attributed to human factors such as weak levees, the failure of the levees resulted from unprecedented weather patterns tied to changing climatic conditions. Weak levees might have been a necessary condition, but that alone would not have mattered if, in August 2005, Hurricane Katrina had not made landfall on the city.

Evidence suggests we are close to a point in the environmental degradation of the planet that is beyond irretrievability except over very long windows of the future (Dow and Downing, 2007). Although there are skeptics, few doubt that the planet is getting hotter by significant amounts or that climatic patterns are shifting enough to cause more catastrophic events (Anderegg et al., 2010). Though at the extreme it is possible, as some believe, that we might be in the early phase of another mass extinction on the planet—this time of the human species—it is likely that we are stressing the ecosystem in ways that generate irreparable harm (McMichael, 1993).

The tragic thing about the climate change debate is that the "truth" matters most to us, not to Earth. If those who support the anthropogenic arguments about climate change are correct, and we do nothing, we are the ones who lose. Other combinations of right and wrong are simply more or less costly. The arguments from the perspective of Earth are rooted in the geological processes that are or are not doing harm. There are risks in being wrong and costs associated with the gamble. But this gamble is hardly just an individual gamble. If we fail to act and we turn out to be wrong, everybody loses, including those who tried to make changes.

By almost all accounts, rising global temperatures, the melting of ice shelves, and dramatic climate variability are taking hold (IPCC, 2014). We understand the causes, the inflection point (550 ppb), and the consequences of our carbon-based lifestyle, so it seems an easy test of human cooperation to prevent our own destruction by finding a cooperative way to reduce our impact on our ecosystem by stabilizing our climate. And yet, given all this, we were unable to generate a binding international agreement or national legislation that could dramatically alter our current trajectory (Bianco et al., 2013). Contemporary political behavior suggests that the required level of cooperation is far from attainable. The goal of this book is to understand this conundrum and to examine more rational ways to implement solutions.

I start with a brief description of the problem itself from a nontechnical perspective. My task is not to present new scientific evidence that supports or refutes a particular understanding of the climate cycle or the role of anthropogenic contributions to the warming planet, but rather to clarify the hurdles we face in the political arena. Resolving many of the environmental problems is not primarily a technical challenge but rather a political one. At the most basic level, reducing our consumption of carbon contributes to reductions in CO_2 emissions, which, in turn, reduce the pressure on the planetary ecosystems, and reducing consumption can result from technological advances. But the foresight to embrace those technological advances, the will to compel their development, or the collective recognition that paying the costs is worthwhile resides in the realm of politics. Technology and geology can explain the physical processes that we confront but not the reasons we choose *not* to confront them. The social, technological, and political aspects all come together at a point that we call policy. But, absent the political will, we cannot achieve policy. It is my contention that the biggest hurdles we face are, in fact, political.

The Carbon Cycle and Global Temperatures

The carbon cycle and its planetary impact were not always political issues. For eons, the natural cycles on planet Earth were able to regulate the transfers of carbon-based emissions released into the atmosphere through various forms of carbon sequestering. Organic plants consumed carbon dioxide from the air and sequestered it in tree trunks, bogs, diamonds, oil, and coal, while the oceans sequestered huge amounts of carbon dioxide in the hardened shells of shellfish (Orr et al., 2005). Atmospheric equilibrium was achieved naturally to balance inputs and sequestering. The planet went through cycles of temperature gradients that brought on ice ages and subsequently melted mile-thick ice shelves in a process that appears to have been at least partially a function of CO_2 concentrations in the upper atmosphere (Whitesell, 2011; Singer and Avery, 2007; Michaels and Bolling, 2009). One of the weaknesses in the arguments against blaming human contributions to global warming is the assumption that nature, having pushed the planet through many cycles over the eons, is still the only force at work. There is considerable evidence to suggest that this is no longer the case (IPCC, 2014; Whitesell, 2011; Orr et al., 2005; Bianco et al., 2013; Olivier et al., 2011).

The advent of human-created fires is one of the first instances of nonnatural releases of carbon from otherwise effectively sequestered sources. Over millennia, tree leaves absorbed carbon emissions and stored it in the wood fibers of the limbs and trunk. Humans subsequently rereleased the carbon into the atmosphere when keeping warm and cooking. Trees and other organic materials rotted and formed bogs that were carved up to warm houses, and coal mines that dot the landscapes of Wyoming, Pennsylvania, and other states throughout the United States and other countries started as leaves that grew on trees that absorbed carbon from the atmosphere. Humans burned coal to keep warm, initially, but eventually to fire the furnaces of industry. The process was relatively harmless because the scale was insignificant. Humans burning fires early in this process did not know about the carbon cycle and the effect of its release into the atmosphere. Early humans, though they began the process of our potential climate problems, were as blissfully ignorant of what they were doing as were the dinosaurs that preceded them at the top of an Earth-centric hierarchy.

These encroachments on the ecosystem gave way to more intense violation of the carbon equilibrium with the industrial revolution, the internal combustion engine, petroleum distillation, economic development, and population growth. The current state of societies that are highly industrialized, are highly consumptive, and have large populations puts to the test Earth's ability to rebalance in time to stave off catastrophic climatic consequences. From this perspective, human development has changed the rate at which carbon is being released from its sequestered locations. Also, at the same time, human development has constrained the natural mechanisms for sequestering carbon on a scale that shifts the heating and cooling patterns of the planet. The equilibrium, most climate scientists say, is out of balance, and the deciding contribution is related to human production of CO_2 (Anderegg et al., 2010). Much of the literature concludes that we are rapidly reaching the point at which Earth's ability to cope will be sufficiently degraded and reversal will be impossible in a time frame useful to its contemporary inhabitants. The overwhelming bulk of this evidence appears to point toward anthropogenic causes (DeSombre, 2007; Dow and Downing, 2007; Anderegg et al., 2010; IPCC, 2014). While this debate rages on between the doomsayers and the naysayers, the locations of carbon continue to shift in ways that are increasingly harmful to the very people caught up in it.

As the average temperature of the earth increases, glaciers can no longer sustain their equilibrium between accumulation and melting; as the ambient temperature of water increases, so too does its volume at a rate tied to the thermal expansion of water. Given the immense volume of water in the oceans and the positioning of many of the world's largest population centers, small changes in this volume can have large and disastrous consequences. Some of the great ice shelves on the Antarctic continent hold enough quantities of frozen water to inundate coastal regions if it were to melt or to break free of its land mooring. The breakup of the Larsen Bice shelf in the Antarctic in 2002 is one example; the rapid melting of the Petermann glacier in Greenland in 2010 is another.

There is significant scientific research linking the warming of the oceans to the more dramatic and destructive weather patterns observed around the world (IPCC, 2014; Hansen et al., 2012). El Niño and La Niña are complex weather patterns caused by changes in deep ocean currents and temperatures. These patterns have intensified in recent decades, dramatically impacting weather patterns in the continental United States. Climate science demonstrates that these are no longer debated or debatable processes. What generates the discord are the questions of whether humans are a large enough contributor to influence the much larger ecosystem and whether the time scale for any changes are influenced by our activities. There is agreement on the fact that a glacier melts when the temperature balance that determines the rate of melting and accumulating ice has tipped toward melting. Disagreements tend to be over whether human behavior is a major contributor, and, if it is, whether it can be a contributor in reversing those changes.

Those who find little reason to accept anthropogenic explanations for global warming and look to long-term patterns of planetary heating and cooling instead take the view that this is a natural process and that the human contribution is so small relative to geological processes that any changes we make to mitigate warming will not pay dividends (e.g., Singer and Avery, 2007; Michaels and Balling, 2009; Michaels, 2003). Their view is that the temperature of the earth has gone through numerous cycles over the past four billion years and what we are facing today could be just another of those cyclical trends that have had little to do with human behavior (Whitesell, 2011). If we were to change our behavior in an effort to influence this warming cycle and the natural cycle continued to occur, the outcome would be the same, but we would have paid a significant cost in terms of social adaptation. Moreover, the retreat of the ice shelves from their maximum extension during the ice ages are greater than those we are experiencing today. Poles were once warm, deserts lush, and tropical regions arid, and all these changes took place without human intervention. But regardless of whether or not ecosystem changes are part of a long-term trend, there is sufficient evidence that human consumptive behavior is shortening the time in which changes are occurring. The precarious positioning of human population centers were obviously not a consideration in the previous climatic cycles either.

It should be clear that I think somewhat like that proverbial soldier in the foxhole—the cost of being wrong on climate change far exceeds the cost of social adaptations that might have little short-term impact. If nature will get us to the point of human extinction in 1,000 or 10,000 years and human behaviors will accelerate that to 400 years, it is in our collective interest to have those extra 600 years to try to turn things around. The point of irreversibility is critical to those who stress the importance of taking steps today in order to ameliorate the long- and short-term consequences. We all go through life evaluating outcomes and estimating the likelihood of one over another. If you lease a car, you sign an agreement that specifies the value of that car two, three, or four years into the future. Assuming natural wear and tear, the price you pay is a function of an estimate of its future value. As we argue about climate change and the anthropogenic contributions—both to the problem and the possible solutions—we are estimating that future value. A small chance of being wrong has a very high cost if we do nothing. It is a cost that dwarfs that of erring on the side of action today for an uncertain outcome in the future.

One solution lies in technological advances, and numerous applications are already available. Lighting accounts for roughly 7 percent of average household electric consumption. LED and CFC lightbulb technology reduces this consumption by up to 75 percent. The only action required is for consumers to change the lightbulbs they use. Unfortunately, there is a lot of discussion over whether or not we should be compelled to do so. So much, in fact, that it seems as though freedom to buy inefficient lightbulbs is more important than the dire consequences of a rapidly warming planet. Many large countries had banned incandescent lightbulbs by 2012, but the United States delayed conversion until 2014, and even then only for wattages in the range of 100 to 40 watts (US Congress, 2007). The situation is no different when it comes to fuel standards. Europe and China have more restrictive regulations on fuel economy than we do in the United States, and it took a unilateral presidential directive to increase U.S. CAFE (Corporate Average Fuel Economy) standard to international levels. In this country, we debate whether we should have a higher fuel economy standard, if any standard at all. It is not technology but politics that constrains our behavior. This leads to the question of why reducing the production of a gas that is wreaking havoc on our planet should be a controversial issue. We all have a collective interest in preserving the viability of our ecosystem, and we have multiple institutional testaments to that collective challenge (IPCC, Kyoto, Copenhagen). Unfortunately, that is not the pathway of politics (Bernauer, 2013).

From the perspective of a human contribution to climate change, the carbon cycle is a function of consumption. Carbon exists on the planet in quantities that are largely stable over time. Humans cannot create carbon; we simply move it around by combining it with other atoms to create molecules such as CO_2. When early humans made fires for warmth or cooking, they were consuming comfort or food. Today, too, we produce carbon dioxide through consumption; we are just more sophisticated about how we do it. Heating in winter and cooling in summer generally work by burning fossil fuels; cooking on a grill or stove is also made possible by carbon-based fuel. Travel uses carbon-based fuel today, whereas early humans used only their own energy; they traveled by walking. The more we travel, the warmer we make our homes, and the more we eat, the more fossil fuels we burn. In fact, we burn fuel to manufacture the car or plane that we use for transportation and to deliver the materials to build our homes that house our stoves that cook our meals. We consume a lot more per capita than early humans ever did. Climate change and global warming are, in part, functions of the human production of CO_2 gases, and our consumption lies at the core of that process. Patterns of consumption, as it turns out, are also a political issue. Technology could and would reduce our production of CO_2; politics can prevent—or at least impede—those technological advances from being used at all.

The natural carbon cycle produces CO_2 just as it did before humans began to occupy the earth. Scientists relate the warming and cooling of the planet over the past few billion years to the carbon cycle, where melting ice is linked to excess CO_2 in the atmosphere and the cold periods to a relative deficit of it (Whitesell, 2011). Geological processes shape the release and sequestering of carbon dioxide, and, as the relative composition of our atmosphere changes, so too does the temperature of the earth. The earth's natural contribution to CO_2 levels in the atmosphere, CO_2 released from trees for example, is not a form of consumption.

As the global population increases, so too does the amount of CO_2 released in the aggregate. Therefore, even if individuals restrict their own consumption, as a group, we continue to increase the amount of CO_2 we release into the atmosphere. The good behavior of any one person, then, is nowhere near enough to change the trend in the anthropogenic contribution to climate change. To make things worse, though some countries are reducing the production of CO_2 on a per capita basis, other countries are just coming online as being major consumers. As a result, we have a long way to go before CO_2 emission begins to reduce as a result of industrial efficiency alone.

Fundamentally, consumption that produces CO_2 emissions is at the heart of anthropogenic causes of climate change. If we could consume less or consume significantly more efficiently, there is a window that allows for a new and sustainable equilibrium. The time horizon between any changes made to consumptive behavior today and its impact on the planet's ecosystem is very long. So long, in fact, that most of us will not live to see the rewards of our behavioral efforts if we were even to make those efforts. Our children or grandchildren might, but not us. Climate models suggest that over the long-term (a century or more) reductions in CO_2 will significantly and positively impact our ecosystem, but little change in climatic conditions should be expected over a shorter term (Hamlet, 2011). A single big storm resulting from global warming will not cause a mass extinction of humans either—those changes, too, will come slowly and gradually, perhaps to the point when we realize we missed the opportunity.

It is hard to look far enough into the future to understand how we might be initializing changes that could make our planet uninhabitable. It might seem counterintuitive, but understanding a problem and its potential consequences does not always lead to behaviors necessary for a solution. It might sometimes be rational for an individual *not* to act in spite of evidence suggesting they should. The collective action problem is primarily responsible for this conundrum. Technology is not the key to overcoming this tendency of individuals to fail to act, nor is fear or altruism. Politics has the ability to encourage or coerce collective action.

Why Can't We Fix This Problem?

This problem is serious, perhaps the most serious problem we face. The causes have been linked to human behavior, and the potential consequences are catastrophic. Other international issues that involved collective goods have been successfully addressed through international coordination, most notably ozone depletion and whaling (DeSombre, 2007; Sandler, 2004). So why is the problem of climate change so intractable? The answer lies in what we know about groups or individuals securing a collective good. Many collective goods have been obtained through the coordinated action of communities, and many of these were once problems that needed solutions. Safe roads, schools, and police protection have all been managed through collective action at the local level, and numerous international agreements that compel individual citizens to alter their behavior to benefit all citizens have been signed and implemented. An example of this is the ozone hole, caused by man-made ozone depleting substances. We no longer use CFCs to cool refrigerators or to propel hairspray and deodorant. The problem was so extreme that nightly news programs in New Zealand in the 1990s began with a graphic depiction of the size of the ozone hole and reminders to wear appropriate clothing and use sun block. Not doing so caused a dramatic increase in the incidence of skin cancer. Individuals might not have realized that the aerosol cap was replaced by the pump spray in a multitude of products, but this comparatively minor change is helping to close the ozone hole over Antarctica.

Our ability to think about a collective response to climate change stems, in part, from our understanding of collective goods. A collective good is a resource from which nobody can be excluded, regardless of whether or not they contribute to securing that good (Olson, 1965). In contrast to private goods, from which an individual who does not contribute to obtaining the good can be excluded, achieving an optimal level of a public good suffers because of the incentive to "free ride." Air quality, weather patterns, and ocean temperatures are goods from which we all benefit equally even though the contribution toward maintaining the viability of these goods is disproportionate. To understand why political solutions to the climate change problem are not on the immediate horizon, we have to consider the conditions under which public goods can be achieved.

From national security, clean water, and clean air to freely accessible radio and television stations, public goods are all around us, and we partake in them every day. The contrast to a public good is a private good, which costs an individual or community a fee. The car you buy, golf club you join, and personal items you choose to consume are private goods.

Two key conditions define a public good. The first is that no one individual can be excluded from acquiring or consuming the good. Clean air is a prime example. Regardless of whether or not someone contributes to keeping air clean, everyone gets to breathe it. In 1970, the U.S. Congress passed the Clean Air Act. Regardless of an individual's opinion on such forms of regulation, everyone benefits from the better air quality. Water is much the same, and national security has many public good attributes. Whether or not an individual supports a strong military or any other component of national security, nobody is excluded from a secure country. A stable climate, too, is a public good from which individuals or communities cannot be excluded.

The second condition that defines a good as public is that the good not be rivaled. A nonrivaled good is one in which consumption by one person does not subtract from consumption by another person. One person breathing clean air has no impact on anyone else breathing it, nor does one person's access to national security diminish any other's. These goods are not consumed in finite increments. Some nonexcludable goods are rivaled, such as fish, trees for timber, and parking spaces close to mall buildings. There are a finite number of fish in public lakes—for every fish caught there is one less for the next person.

In order for our climate to have a chance to stabilize, people who cannot be excluded from this good must reduce the types of consumption that cause climate change. This is not easy to achieve, because individuals have the incentive to consume as they have been without worrying about the quality of the air or the stability of the climate. Their contribution to the problem is so small relative to its magnitude that the solution does not appear to be in their hands. Moreover, since any one individual's impact is so small and the numbers of people making this contribution are so huge, it is hard to observe when someone defects from the cooperative effort. In fact, it is quite rational to defect, even if it seems counterintuitive to do so.

Elinor Ostrom (1999; and with Dietz and Stern, 2003) has described mechanisms for the self-regulation of a common pool resource, and for her body of work on this topic, she was awarded the Noble Prize in Economics in 2009. A common pool resource is a form of a public good that is nonexcludable but rivaled—individual access cannot be denied, but for each unit consumed, one less unit of that good remains. The stability of our fish stocks and the overfishing of certain habitats have bedeviled public policy for some time. Cod fisheries in Canada collapsed from overfishing, and fishing had to be banned or catch limits imposed, as did anchovy fisheries in Peru in the late 1970s, which have struggled to recover ever since. Self-regulation requires an individual incentive to overcome the tragedy of overexploitation, and much of Ostrom's work is on the regulation of fisheries and other of contestable public resources. Fisheries represent a rather small-scale collective problem relative to global climatic changes, but the core principles are the same. Each individual has incentive to exploit the common pool of resources in spite of the potentially deleterious consequences of doing so. Solving collective action problems is not easy, but Ostrom has pointed to possible solutions that rely on self-regulating cooperative behaviors.

Mancur Olson (1965) took a different path to understanding collective behavior in social problems related to public goods. To Olson, the analogy of behavior of the firm in a market provides the operative logic behind understanding the achievement of collective goods. Importantly, Olson sees a difference in the optimal achievement of the goods based on the size of the group attempting to maximize output. Olson's argument, in short, says that when the group size is large, any individual defection from the contribution toward achieving a public good will go unnoticed and have only a minor impact on the preferred outcome. And since each individual calculates that there is little cost to defection, they defect. When all individuals avoid the cost of providing for a public good, the group achieves at best a suboptimal outcome, even though all would prefer to secure the good in question. To put this into concrete terms, to get a stable climate requires reducing the amount of CO_2 we put into the air, and each trip to the store produces tailpipe emissions that contribute to the amount of atmospheric CO_2. So in theory, walking to the store or taking public transportation would improve our chances of climate stability, but each individual contribution from that one trip to the store is so small relative to the magnitude of the problem that walking might not be rational. Moreover, so many must contribute that any one trip to the store gets lost in the mass of people and the numerous other trips they take. Each individual could reasonably conclude that a single trip will not affect global temperatures. And since each individual can come to that same conclusion, there is no change in our collective behavior. We all take small trips in cars, and in the aggregate, we continue to degrade our global ecosystem. The key to Olson's argument is the marginal contribution of each individual's interaction with the size of the group trying to optimize its outcome. One of the differences between the work and inferences of Ostrom and Olson is the assumed size of the group. Ostrom's problem of the tragedy of the commons revolves around common pool resources that are enjoyed by relatively small groups over contestable resources. Olson sees group size as one of the main determinants of optimizing outcomes.

Behaviors that affect ecosystems at the planetary scale involve large groups. The best way to tackle the difficulty of securing policies to reduce greenhouse-causing emissions is in terms of the collective action problems articulated by Olson. But central to his argument are the identity and capabilities of the actors most influential in their ability to generate climate change legislation or regulation. Because of the scale of the problem of climate change, no one individual can influence climate patterns by reducing his or her consumption of CO_2-generating behaviors, what is a wide ranging collective response, but oddly, one that no one individual has incentive to participate in.

A Theory of How This Works

Reducing levels of CO_2 emissions enough to stem the consequences of climate change involves group interactions from at least three levels: the individual, the country, and the international community. The individual level is critical at the point of consumption. Reduction in the production of greenhouse gases cannot occur without reductions in consumption at the individual or industrial level. Since industry produces for the individual, individuals have to consume less of the products that cause carbon emissions or consume in ways that produce less carbon emissions.

At the national level, legislation is required. National legislation can entice or compel restrictions in consumption, and legislatures that pass laws restricting consumption are a group that must solve the collective action problem. Legislatures have to maximize output given competing preferences and the ensuing compromises. And because remedial action is required of all carbon emitting countries—although at different levels of commitment—group interaction is needed among states in the international system. This interaction would involve treaties that dictate compliance at the state level. An international treaty to reduce CO_2 emissions might compel a national legislature to enact laws that restrict consumption of products that generate too much CO_2.

One example might be CAFE standards for automobiles, in which a country would enact legislation that requires cars to achieve a certain standard of fuel efficiency. These national standards, in turn, restrict the type of automobiles manufactured and subsequently the range of cars an individual can buy. The international treaty does not mandate CAFE standards, only that the country find a way to reduce CO_2 emissions. The national legislature or executive decisions translate the demands of the treaty into policy. Policy, in turn, provides incentives or constraints on the individual who can act in ways consistent with reaching the targets demanded by treaty.

At each level, group sizes are significantly different, providing a way to think about group size as the mechanism that accounts for the efficient use of common resources. These different group sizes are precisely the loci where Olson suggests we will find differing levels of efficiency in the production of a public good, and as Ostrom points out, smaller groups should be able to self-regulate to maximize the provision of common resource pool goods. Most importantly, these group actions are not independent. That is, if a country signs an international climate treaty that compels restriction of products that emit greenhouse gasses, individuals have to conserve the most. The legislature, then, must turn international treaty into domestic laws that demand or entice compliance by individuals. These three principle levels are made up of three differently sized groups who each have different identities and who all have to coordinate in some way to overcome the problem of collective action.

Some argue that technology will save the planet, and that technological adaptation will be driven by market forces. This may be right, but constraints that help make the technological solutions economically viable may also be required. Given a choice to buy a product that used less energy and therefore generated lower CO_2 emissions but cost considerably more than those currently available, an individual would have no reason to opt for something more expensive that did the same thing. Altruism might be one reason, but argument and evidence can quickly dismiss the large-scale remedial implications of a few altruistic consumers. Another reason might be the cost of electricity. In this case, the consumer must have a long-window view of the future, because a small monthly saving in electricity costs would have to offset the immediate outlay of extra funds. There are altruistic individuals and also some who look to the future and demand energy-efficient products. Industry, however, has figured out that there is not enough demand for energy-efficient products and therefore does not make them available to the average consumer (Rosenthal, 2011).

One such product is the DVR. American consumers prefer that their unit turn on immediately. To achieve this, the hard drive continues running even when the DVR and TV are off. By EPA accounts, DVR players use as much or more energy in a year than the average refrigerator. Energy Star ratings are available for refrigerators but not DVR players, and service providers do not make an energy-efficient DVR option available because there is little or no individual demand for them to do so. Technology cannot be the panacea without a change in individual behavior, and individuals do not always make choices that will take us toward a stable planet. Because there is no demand, no national legislation requires a change in DVR products. An international treaty that compelled reductions in the production of CO_2 might lead to legislation that required energy-efficient DVR players. This is a behavioral change that could be forced by international treaty.

In functional terms, reductions in CO_2 emissions have to come through reductions in individual and industrial consumption of carbon-based products. Governments consume, but not at the level of individuals or industry. Since the incentive to free ride is strong, individual-level behavior will, in the norm, have to be coerced or incentivized through government legislation. Constraints can come in the form of selective incentives for those who consume less or legislative mandates proscribing certain behaviors. Because individual- and national-level action can be financially and politically costly, few states will significantly constrain individual-level consumption absent an international treaty that provides for collective contributions. That is, most states do not, by themselves, have the ability to affect climate change patterns, and those that do will be reluctant to act independently, just as any one individual might be reluctant to act. The immediate political implications are evident in the rancor over promoting "green jobs" or "green energy." One view is that this is an economic disaster ruining competitiveness and ensuring economic decline. The other view is that it is a necessary step forward that will create jobs and help with environmental issues (Broder, 2012).

Olson's theory (1965)[1] relates a group's size and composition to the ability to provide a collective good. Since every individual has an incentive to free ride on the cost of providing nonexcludable goods, the larger the group size, the less likely a defection will be noticed, or punished if noticed, and ultimately, the less likely that a large group will provide for an efficient amount of the good. To make Oslon's argument concrete, imagine a group of 10,000 people who have to coordinate to achieve some socially optimal outcome, such as a stable climate. If any one person decided against participation in the project, 9,999 people could still see it through. Most of these would not even notice one absence. The larger the group, the easier defection is. Now, if the group contained only 100 people, defecting would still be relatively easy, but the smaller the group, the more obvious the defection, and the greater the implications for the completion of the project. Olson's argument lays out the foundations for why large groups have a lot of defectors—or free riders—and are therefore less likely to meet their goals, while small groups are more likely to be successful. He has developed a mathematical explanation for why this happens, which, in abbreviated detail, forms the backbone of my argument.

The core of Olson's argument is that the rate of gain to the group must exceed the rate of increase in the cost of providing the good, and do so by at least the same margin that an individual gains from the provision of the good. In other words, if it costs more to produce the good than either the group or individual get from having the good, few will pay the cost, and the public good will fall short. Not many people would voluntarily pay more for fish so that a fisherman could catch less and still make a living. But by taking the pressure off the fishing fleet to maximize their catch, we would help ensure viable

1. Olson's argument was developed further by Sander (2004). Sandler adopts the term "socially optimal," which provides a much clearer understanding of the potential outcome from collective action problems; I rely on his terminology of social optimality.

fisheries. As with beef or dairy products, the true cost of producing without subsidy or cartelization of the industry is too high. Much of the U.S. agricultural sector is subsidized so that the farming community can survive—at least the family farm—and without these subsidies consumers would have to pay the true cost of production (Congressional Budget Office, 2014). Similarly, there is a reluctance at the individual level to pay the cost of forms of consumption that might help stabilize our climate.

In order to demonstrate how this argument holds, I am going to express it mathematically to break down the critical components of the argument so that we can see how they interact. For example, there is some fraction of a good that accrues to the individual, F_i. This might be thought of in terms of the individual's value for a stable climate, a public park, or public welfare, and it can be thought of in terms of the value to the individual relative to the value to the group, V_i / V_g, where V_i and V_g are the value of the good to the individual and group, respectively. Here, we might think of this as each individual's value for a stable climate relative to the United States' value or the state or city in which the individual lives. There is also a cost of achieving that good, C. For individuals to contribute to the achievement of the public good, they have to gain more than it costs. For climate change issues, this might reflect the relative cost of an efficient DVR, the cost of an efficient car, the difference in cost between incandescent lightbulbs and LED lights, or the added cost of gasoline that results from a tax to reduce consumption. If the payoff from any of these products is less than their costs, it is rational to buy the cheapest even if it consumes more energy and produces more CO_2. If individuals spend more than they get back in value, an economist will generally see that as nonrational behavior; so too might the individual.

According to Olson, for individuals acting rationally, the public good will be provided if $F_i > C / V_g$, where $F_i = V_i / V_g$ and by extension, $V_i > C$. In other words, individuals will participate in the production of a collective good when the value of the shares they expect to get are greater than the cost they pay to acquire the good. If the conditions are not met, each individual has an incentive to free ride. In a large group, the share to any one individual is small, and there tends to be less of a payoff to the individual. This requires, for the participation of the individual, a cost of participating that is "so small in relation to the gain of the group . . . that the total gain exceeds the total costs" for the individual by greater margins than the gains accrue to the group (Olson, 1965, p. 24). As Todd Sandler frames this, the benefits of the collective good minus the costs required to secure it must be greater than zero. Since an individual gaining more from the public good of climate stability than the group as a whole resides predominantly in the domain of psychology, it is unlikely that any one individual will pay the cost of the provision of the public good. By this logic, relying solely on the behavior of individuals to act in their own interests is not workable (2004, p. 24). The problem of climate change is simply too big relative to the contribution of any individual.

However, compliance by the individual might not reflect the preferences of that individual from a strictly rational calculation, and in fact there is argument and evidence to suggest that sometimes individuals participate based on culturally derived norms or socially dominant standards (Chong, 2000 and 1991). Free riding might not be the preferred strategy, but given perceptions of social acceptability or wider compliance, individuals will reluctantly defect. Individuals might wish to work toward a world where the climate is not a threat to human viability, but our social norms look askance at those who act in an altruistic manner. "The liberal," "the tree hugger," and "the environmentalist" are labels that tend to set this group outside the mainstream. This is evident in market behaviors required at the individual level to reverse trends in carbon-based emissions. For example, sales of low-energy lightbulbs, fuel-efficient automobiles, and non-carbon-based electricity production have been meager relative to the more traditional forms of lighting, transportation, and energy.

There are many who adopt technologies that produce less carbon emissions without recourse to selective incentives or coercion by government regulation, but free riders remain the dominant actors. People who pay the added cost of buying the new technology are called early adopters. They might do so because they want to be at the cutting edge of technology, or because it contributes to the achievement of a collective good. But early adopters tend not to act in a strictly rational way, at least in the way economists consider rationality—maximizing benefits against costs. Buyers of the Toyota Prius and Chevy Volt are great examples. The Prius introduced hybrid technology and had low early sales numbers, and the cost was relatively high. The weekly gas savings to the consumer were low, as was the contribution to the solution to climate change (U.S. DoE, 2014). Today, however, the price of a Prius is comparable to other models, and volume of sales rivals that of other high-volume cars. Early adopters provided the mechanism to generate economies of scale and turn the Prius from a novelty into a mass-produced vehicle. Sales of the Volt started off rather sluggishly too, and the cost of buying early was relatively high (U.S. DoE, 2014). But in terms of the production of CO_2 gases, the Volt is an improvement on traditional cars. Those advocating a strictly market-based explanation for this would be misled. Government subsidies to the tune of about $3,500 per Prius were necessary to get those early adopters to purchase the car; a similar incentive structure is motivating sales of the Volt. So the market alone, or the existence of the technology alone, was insufficient to get people to act counter to their rational interests. We, as a national group, agreed to pay some of the cost of these early vehicles so that we could reduce our collective production of CO_2 gases and generate a market that could achieve economies of scale.

Prisoner's dilemma is one theoretical tool for thinking about why we, as individuals, balk at voluntarily contributing to a collective good, or, to turn that around, why it makes sense for us to free ride on the efforts of others. This theoretical device allows us to think about the conditions under which it is rational to cooperate rather than defect. Collective action problems can be seen as a prisoner's dilemma where the rational move is to defect (Sandler, 2004, pp. 23–27). What this means is that the payoff for "winning" from your neighbor or group has more value than cooperating. A normal prisoner's dilemma is modeled using a two-person scenario, but the logic would hold under conditions of an *n*-person prisoner's dilemma, which we can think of in terms of groups. Sandler makes clear, however, that not all collective action problems are best represented by a prisoner's dilemma. To overcome the collective action problem in an *n*-person prisoner's dilemma, costs could be imposed on those who defect that are equal to the difference between the costs and benefits of voluntary compliance. But somebody has to impose those costs, which can be a way to compel compliance with actions that promote the public good. An alternative framework for thinking about collective action problems is what is referred to as an assurance game, where one of the equilibria provides for a socially optimum outcome. That is, we can think about these rational games with a pathway to cooperation. The assurance game framework provides for a dominant strategy that results from coordination between actors, particularly if one actor takes a leadership role. For example, the leading actor can bind themselves to an agreement by committing early to an outcome or committing to punish defectors (Sandler, 2004). In effect, how we view future payoffs relative to the immediate costs is a crucial component of the dilemmas we face.

If we can accept that individuals are reluctant to provide a collective good without enticement or coercion because it is too easy to defect, understanding the collective reluctance to change our consumption patterns requires that we ask why national and international groups do not or cannot coerce, constrain, or cajole individuals to overcome the incentive to free ride. If they could or would do this, individuals would contribute to the solutions to the problem of climate change. This is what politics is all about. Climate change is a collective problem that requires action from the largest possible group—individuals—that we know will not, of its own accord, act for the optimal outcome. An international treaty that compels states to manipulate the consumption of industry and individuals in a way that reduces the impact on the planet would be the ideal solution. It has long been understood that the solution to the collective action problem lies in selective benefits for participation (Lichbach, 1994), and absent an individual motive to contribute based on cultural, sociological, or other factors (Chong, 2000), states have an incentive to generate sufficient levels of public goods. If we accept individual compliance under a narrow set of nonincentivized or noncoerced conditions, but also accept that this level of participation is insufficient to redress issues of climate change, understanding the politics behind national and international behavior is critical.

Olson's argument speaks to the interactions within smaller groups with strong leaders or with oligopolistic actors with sufficient interest in providing an optimal level of the public good. The combination of how large the group is and who its members are forms the backbone of his logic of collective action and points to the ability to provide all or some of the collective good. For example, if some subset of a larger group is willing to provide the public good at its own expense, Olson suggests that the subset might provide some but not all of the good necessary (1965, p. 27). When it comes to CO_2 emissions and their impact on climate patterns, a number of subsets of the population might be willing to pay higher costs, act as enforcer, or otherwise coordinate actions in pursuit of a public good. In theory, this should provide us with a modicum of confidence that international or national legislation will address climate change issues. But with hydrocarbon emissions, there is a threshold beyond which, for any given additional effort, there is not an optimal outcome. Many scientists see this threshold residing somewhere around 550 ppm of CO_2 in the atmosphere; others see mitigation as primarily a long-term (a century) solution, and that over the coming decades, regardless of our immediate actions, CO_2 concentrations will increase (Hamlet, 2011; Solomon et al., 2007). From this vantage point, we are playing to a long-term strategy for mitigation and a short-term one for adaptation. Individual-level group compliance, then, requires a national and international commitment. The key is in understanding the behaviors and political motivations of these decidedly smaller groups.

An Extension of the Model

Nongovernmental and intergovernmental organizations (NGOs and IGOs) are national and international groups whose missions are to advance environmental security or deal specifically with climate change and populate and influence this process. The Intergovernmental Panel on Climate Change (IPCC) is one such IGO that brings together the evidence on the questions and consequences of climatic changes (DeSombre, 2007; Bernauer, 2013). The IPCC won the Nobel Peace Prize in 2007 for its work. These organizations exist at the international, national, and local levels, but by and large, they tend to be advisory to the political process. In this role, they can inform the debate, shape policy, and educate the public, but their main mission is to get politicians to accept their arguments. The willingness to do that is often influenced by the politics of the day. These organizations do influence the political climate enough to get policy enacted, and the results are evident in the clean air and clean water acts, the work of the Environmental Protection Agency, the negotiation of the Kyoto climate treaty, and most recently, the president's requirement that CAFE standards for U.S. cars reach 54 miles per gallon by the year 2025. The politics behind this are complex, and it is a long way from negotiating a treaty to getting it ratified in the United States (and other countries) and abiding by the terms of the agreement. Thomas Bernauer (2013) asks why global problem solving is so difficult. Much of the answer lies in the complexity of the politics at each level. Extending a model of collective action should provide some clarity.

According to Olson's logic, the ability to secure a collective good, even among members of smaller groups, will be suboptimal and equal to the amount that the largest contributor is willing to provide in pursuit of that good. Groups with members of unequal size or capabilities will tend toward a suboptimal outcome that is determined by the contribution of the largest member(s). In terms of an international initiative to address climate change, the United States and China and a few more of the largest producers of CO_2 emissions have to improve, but even when they do, each will be limited in how far it will go toward contributing to the solution. Neither the United States nor China wants to contribute more to resolving climate issues as a result of a successful international effort than it gets back. No country wants to lose relative to another, as is evident from the way political aspirants talk about jobs generated or lost by promoting green technology. The cost of contributing toward a public good would not be in proportion to the benefits accrued in trying to get there, with the largest member shouldering a disproportionate share of the cost (Olson, 1965, p. 29; Sandler, 2004). So the wealthiest and the largest contributors to our climate problems will pay the most for the solution, and this is exactly what theory would tell us should happen. Since the potential consequences are so grave, not taking action to reduce CO_2 emissions seems to be completely irrational. Quite the opposite is true, but this example illustrates the difficulty of the collective problem we face.

The foundational arguments for thinking about participation in pursuit of public goods can be summed up in Olson's words, "In any group in which participation is *voluntary*, the member or members whose shares of the marginal cost exceed their shares of the additional benefits will stop contributing to the achievement of the collective good *before* the group optimum has been reached" (1965, p. 31, italics in original). One key word is *voluntary*, and it begs the question of whether nonvoluntary compliance to reduce CO_2 emissions can work to stave off a climatic catastrophe. In many ways, however, Olson's arguments are too simple relative to the problems caused by CO_2 emissions. Two complications make expanding on these arguments critical to understanding the difficulty in generating agreement to redress behavior and policy.

The first is that maximization confronts three levels of actors, each of which embodies the tendency to produce at suboptimal outcomes, and importantly, the ability to provide for some public good at each level is dependent on the ability to produce at other levels of interaction. If an individual in a group that consists of a country's population has an incentive to free ride on efforts to ensure the planet's survival, then so too does the U.S. congressperson, one individual in the population of representatives, who casts a single vote. The congressperson is a member of the larger group of Congress and also a representative of a district with characteristics that generate incentives to either resist policy change or to free ride on legislative initiatives. Since climate change has a global constituency, each country is one in a group of countries trying to achieve an efficient outcome to a common problem. As we move from the individual citizen of a country to the individual state in the global system, the size of the group declines, leading to an increased potential to achieve some positive outcome, even if not an optimal one. But these groups are not independent. Olson's argument falls short in thinking about the independence among groups and the impact of each on the ability to provide for the collective good. The ability or willingness of the U.S. president to negotiate, sign, and get a treaty to restrict CO_2 production ratified is a function of the political support of Congress, specifically the Senate, which ratifies treaties, and indirectly, individuals in the country. Without public support, it is difficult for the Senate to ratify a treaty, and without an expectation of ratification, it is risky for the president to sign a treaty, and given U.S. history with the Kyoto treaty, without a chance of ratification other countries might be less likely to agree to sign a new agreement. This two-level process of low support leading to a low expectation of ratification occurred in Copenhagen in 2009. It should be clear that these groups that need to maximize their outcomes at their respective levels are also intertwined in the complexity of politics.

The second complication involves the role of time, or as DeSombre (2007) frames it, the discount factor for the future achievement of the good that is sought after. One of the difficulties of securing domestic or international cooperation on climate change initiatives is that observable implications of the problem and those that might result from prospective solutions are temporally distant from the costs incurred in pursuit of CO_2 reductions (IPCC, 2014; Hamlet, 2011). As the consequences of the actions are pushed further into the future, actors—individuals, congressional representatives, and presidents—are more likely to discount future benefits relative to the immediate costs of pursuing those outcomes. When the public good is a sustainable planetary ecosystem and the cost to be paid involves human adaptation and carbon mitigation, there is also a substantial role played by uncertainty. Time magnifies uncertainty. The pursuit of climate change legislation generally works from the assumption that anthropogenic behaviors cause atmospheric accumulation of greenhouse gases at unsustainable rates that ultimately lead to global warming, which in turn alter climate patterns. Climatic changes will generate shifts in habitation and agriculture patterns and have the potential to alter our current relationships with the planet. If climate change is not the result of anthropogenic causes, the cost required to stabilize global temperatures will likely fail to achieve the objectives in spite of the most concerted efforts to do so (Singer and Avery, 2007). The doubt created by arguments that planetary temperatures are part of a long-term cycle influences the willingness of individuals to adapt their behaviors to meet future expectations. It is easier to free ride when uncertainty about cause and effect exists. And it is easier to give this uncertainty greater importance when the time horizon under which the most deleterious consequences of climatic change is far in the future.

The political question, however, is not whether empirical evidence is or is not conclusive, but rather whether a group member perceives it to be so.[2] Creating the perception of uncertainty over the anthropogenic causes of climate change can reflect a motivated bias in either the science or the interpretation of it, or both. The perception of uncertainty, that is, can be driven by an interest in holding a large discount for future benefits from climate remediation. If the costs of change are too high relative to future payoffs, uncertainty over cause and effect are easier to countenance and/or manufacture. Group members who bear a disproportionate share of the costs of climate change legislation are more likely to be skeptical of science that identifies anthropogenic causes. Unsurprisingly, there is considerably more resistance to initiatives to tackle the problem in states where coal and oil are a significant component of the economy. This poses a problem for making the political progress required to get the United States and global community to codify and implement environmental policy.

To put this into Olson's framework, where some of a public good will be provided when the individual benefits more than the group and more than they expect to pay, the expected value of the outcome must be weighted by uncertainty, and the discount factor with which the future is valued relative to the current costs. For example, an individual making an investment in a retirement fund, starting a new business, or buying a new home would think about the possible returns and likely risks involved. Part of their calculating returns involves relative investments—different funds carry more or less risk and pay

2. On the question of whether or not evidence supports the claim that climate change processes are generated by anthropogenic causes, see Anderegg et al. (2010), Roosevelt (2011), and Schneider and Kuntz-Durisetid (2002).

greater or lesser returns—and knowing how long they would stay with the investment. Every LED or CFL lightbulb package provides a calculation of the payoff over time relative to traditional incandescent bulbs, as do EPA stickers on windows of new cars. We are given the information, so we can calculate our current expenditures and our return on those costs.

If we think of the cost to future value ratio as an expected value E(X), then E(X) is a function of both the value of X, the uncertainty of obtaining it, p, and the discount factor as a function of time, q, such that $E(X) = ((V_i * q_i * p_i) - C_i)$, where q and p are in the range of $0 < p$ or $q < 1$. Essentially, we place a value (V) on some outcome, say a stable climate, and we recognize that we have to pay a cost (C) to get there. But our payoff is not so certain (p) because it is not clear that the cost we pay will generate the benefit we anticipate. Also, we value some things more in the present than we do in the future (q). When we are certain about cause and effect, and have a discount rate of climate legislation that is equal to the current value ($q = 1$; $p = 1$), achieving an optimal amount of the public good is only subject to Olson's rational free riding dynamic [$E(X) = V_i - C_i$]. But when individuals discount the future payoff of climate change legislation, their willingness to pay current costs diminishes (Bernauer, 2013; DeSombre, 2007), such that at completely discounted future benefits and complete uncertainty, there is no expected payoff from legislation ($E(X) <= 0$). Even for those who accept that the causes of climate change are anthropogenic ($p = 1$), with perfect remedial policies in place, the pace of change in climate dynamics will push benefits into the distant future (IPCC, 2014; Hamlet, 2011; Solomon, et al., 2007). If people discount the future benefit relative to today's cost, remedial policies will be difficult to achieve ($q < 1$). For most individuals, the payoff from compliance with change-inducing behavior is discounted even though the required costs are immediate. Uncertainty and future discount are linked, as can be seen in a hearing by a committee of the U.S. House of Representatives in 2011 on a bill that would legislate the interpretation of the science behind carbon-based understandings of climatic changes to be "inconclusive" (Broder, 2011; U.S. House of Representatives, 2011), in effect, legislating the value of q in the calculation of the expected payoff from remedial actions to be less than one. From a political perspective, this focuses on whether burning more coal preserves or creates jobs versus the job creation aspect of moving toward green technologies (Broder, 2012). The core element is the interpretation of q and the use of uncertainty, p, to generate support for the alternative positions.

The Path Forward

To sum up this argument, we have a situation where most scientists see the earth as embattled because of the large-scale changes in climatic patterns that result from human consumption that produces CO_2 emissions. The value—or the collective good—we seek to achieve is a stable climate in which the human species can thrive. Given the ability of Earth's ecosystems to adapt, the difficulty in crafting solutions that can be implemented do not solely confront technological hurdles but also significant political ones. Left to scientists and engineers, and without regard for cost or political displacements, the problem at hand would be easy to solve. But this is not how the world works. The main obstacles are political, and in such a political world, we have to attempt to achieve a good (a stable climate) for which we all share equally but have to pay for disproportionately and under conditions where we each have an incentive to rationally avoid paying the cost of reducing our consumption or adapting to avoid future consequences.

In this mix of rational self-interested behavior, it turns out that if we discount the future value of a stable climate relative to what it would cost us today to work to mitigate or adapt to climate change, we have even less incentive to bear the burden of change. And because our understanding of climate change relies on data and processes that were generated long before human habitation of the planet, our science is not completely certain about cause and effect (IPCC, 2014). Uncertainty can be exploited for immediate gain by those who do not want to pay the cost required to make the necessary changes. Although politics is at the core, empirically demonstrating its pernicious role is hard.

The remainder of this book will advance these ideas by first providing evidence to suggest that the collective action problem confronts biases motivated by the cost of compliance with regulation that might benefit us all but put a disproportionate share of the burden on a smaller portion of society. For this, I'll use a vote in the House of Representatives on climate change and the follow-up Copenhagen summit attended by President Obama. This did not work out so well for the prospects of shifting consumption away from CO_2-producing goods. But not all is lost. Two decades ago, science identified a growing ozone hole above the Antarctic and attributed it to the widespread use of CFCs, the gas used as propellants in aerosol cans, refrigerant in air-conditioning units and refrigerators, and a number of other industrial and consumer applications. The depletion of the ozone layer that protected us from the harmful rays from the sun was linked to an increase in skin cancers. We now have international treaties that regulate or ban the use of CFC gasses, and although the ozone layer is not completely repaired, we have made progress in stabilizing ozone levels, and there are indicators that our atmosphere is rebuilding this protective cover.

Oddly enough, these types of environmental success stories are as prevalent as the failures. At one point in the history of the United States, we hunted the wolves of the Great Lakes and Northwest to near extinction. When just a few breeding pairs were left in Yellowstone Park, they were captured and taken to Alaska. One of the unsuspected outcomes of removing the wolf from the ecosystem in Yellowstone is that the deer and the caribou lost a natural predator. Photographs of river basins in Yellowstone Park before the near extinction and removal of the wolf show a vibrant ecosystem with full foliage on the trees. When humans took the wolves away, the deer and caribou didn't have to run for their lives. Within a human generation, the river basins became barren because of overgrazing. The deer and caribou killed the ecosystem of the river basins, and when wolves were reintroduced into the park, the river basins began to thrive again. It was a cooperative arrangement to save wolves from extinction and repopulate the park. Not everyone is happy, and ranchers feel pressure on their stock, but the ecosystem is revitalized because national legislation that provided for a public good was able to constrain the behavior of private individuals and businesses. I will expand on these themes and develop a way to think about how politics can be used to generate a socially optimal outcome even when it confronts vexing social problems.

2

♦ ♦ ♦

COPENHAGEN

THE CLIMATE CHANGE SUMMIT

The real test of the politics behind climate remediation legislation is not found at the level of the individual. There may be data on individual level consumption of electricity and automobile purchases, but these data would provide only numbers, not reasons why individuals consume the way they do. The individual is critical to successfully confronting the challenges posed by climate upheaval, so understanding when and why consumers buy green or conserve is important. Altruism, constrained options, and incentives are some of the possible reasons. For example, the government provides a rebate of $7,500 to individuals that buy a Chevy Volt. And as of 2013, General Motors had sold about 50,000 Volts in the United States. If the number of people who would have purchased the Volt without the government rebate is very small, then the public policy of rebates is critical to pushing consumers toward green purchases. In most instances, uncovering the motivation behind such choices is impossible, so the possible explanation is untestable. Occasionally, though, we do get a glimpse into the process.

In June of 2009, the U.S. House of Representatives passed, by a narrow margin, the American Clean Energy and Security Act (HR

2454) to mandate the reduction of U.S. emissions of CO_2 by 18 percent over the next decade. Some of the bill's supporters included the Sierra Club, the Union of Concerned Scientists, General Electric, and the United Steel Workers. It was opposed by a smaller number of, albeit influential, groups comprised of Greenpeace, the National Mining Association, and the American Petroleum Institute. The makeup of the House at that time gave the president's party about a 75-vote majority; his party also held nearly a 20-vote majority in the Senate. In December of that year, the leaders of 114 countries met in Copenhagen, Denmark, to negotiate an agreement by which carbon-based emissions could be reduced by the 50 percent required to stabilize the ecosystem. Many of the world's leaders came out in favor of this 50 percent targeted reduction, as did President Obama. The dominant player had the outlines for prior approval of an agreement setting a standard of an 18 percent reduction in carbon-based emissions. Brazil, India, Japan, and China had made public statements calling for reductions much more significant than the 18 percent mandated by HR 2454 (Dimitrov, 2010). It is puzzling, then, that the Copenhagen summit failed. China and the United States could not coordinate to generate an agreement, and a reasonably small group of national leaders led by a smaller group with oligopoly interests walked away virtually empty-handed. In spite of the optimism that drew President Obama to the summit, the outcome was suboptimal.

These two processes—one within the U.S. Congress and one at the international summit—represent two groups with different numbers of central actors acting on different sets of incentives. The United States was the largest player at the Copenhagen summit and, at least in theory, had the ability to coordinate a response and, again in theory, had the most interest in doing so. However, in spite of the House approval of HR 2454, no binding agreement was reached at Copenhagen, no target emissions were set, and only a vague understanding to limit global temperature increases to less than 2°C over the next decade was articulated. Copenhagen was, by many accounts, a failure cloaked in platitudes and did not come close to achieving its goals. With all the good intentions, with all the resources spent in getting to that point, with the background of Kyoto and the

Intergovernmental Panel on Climate Change (IPCC) reports, and knowing the consequences of inaction, self-interested political incentives stopped this group from securing the public good.

The failure at Copenhagen is puzzling for three reasons: 1) the dominant player, the United States, had something of a prior approval for an agreement reducing carbon-based emissions by 18 percent, 2) China and the United States could not find a way to coordinate an agreement that would provide some modicum of a successful treaty, and 3) a small group of national leaders walked away virtually empty-handed in spite of the consequences. The logic of collective action might suggest that an agreement somewhat consistent with the congressional limits spelled out in HR 2454 coordinated by China, the United States, Brazil, and a handful of other large countries would have been achieved at Copenhagen.

The terms of international treaties are negotiated by the president with his international partners and then ratified by the Senate. Many other countries have similar institutional constraints on their chief executive. This two-step process requires that the president negotiate with an eye to the Senate and, in doing so, coordinate with the leadership in the Senate as negotiations move forward. Sometimes, negotiations might extend beyond what the president might expect the Senate to ratify, and in such instances the president must walk away from the negotiations without a treaty, compel changes to the treaty that comport with Senate demands, or coerce enough senators to vote for ratification in spite of their reservations (Putnam, 1988). The president can also sign a treaty for which he fails to secure ratification and then choose to abide by the terms of the treaty in spite of the lack of congressional approval. The recent arms control agreement with Russia is a contemporary example, as is the Panama Canal Treaty, which the president signed in spite of congressional resistance and successfully got congressional ratification in both instances (U.S. Department of State, 2014).

Treaties are generally approved by the U.S. Senate. Out of roughly 1,600 treaties submitted by the president to the Senate for advice and consent, 21 were rejected by vote, and 85 were never taken up by the Senate and therefore lingered there without confirmation. Given this historical record, the odds of ratification by the Senate of any given treaty is about 95 percent (U.S. Senate, 2014). The negotiations at Copenhagen never got to the point of being submitted to the Senate for ratification.

For Copenhagen to be successful, President Obama would have had to agree to its terms, sign the treaty, and then get the Senate to ratify it. He had a 59-vote majority—enough to easily generate a majority vote and nearly enough to prevent a filibuster. Among the differences, as we will see later, between arms control, the Panama Canal, elimination of CFCs, and reintroducing wolves on one hand, and something as complex and comprehensive as reducing CO_2 emissions on the other, are the interests of domestic constituencies. That is, the three groups are engaged in trying to maximize their self-interests and the public good at the same time. Olson's logic will tell us that self-interest wins when the group is large, but that public good could win in a smaller group. Copenhagen had a chance, but a slim one.

The Copenhagen negotiations were complex, and multiple factors went into their tone and content. It is, of course, an oversimplification to say that the treaty fell apart because there was only a very slim chance that the U.S. Senate would ratify any treaty that came out of the negotiations. And even though a treaty was not signed, Copenhagen did generate the recognition that something must be done at both the national and individual levels. Kyoto had already codified this recognition, and the dialogue at Copenhagen reinforced it. But Copenhagen did not set mandated and enforceable limits, nor even targeted reductions in emissions. I focus exclusively on the role of self-interested behavior and the two-level interactions to illuminate the political processes and interests that underlie the politics of climate change. Copenhagen is one step on that long ladder, and it allows us to generate an empirical test that can clarify the implications of that political process.

Collective action and multilevel decision-making provide a foundation for thinking about the negotiation and ratification of an international treaty on climate change that could have come from the Copenhagen summit. Since a treaty was never signed, it remains counterfactual as to whether the Senate would have ratified any such treaty, but we can get a sense of the politics behind the potential ratification by looking at evidence from the House votes for HR 2454. Using certain assumptions about senatorial preferences for a climate treaty based on congressional district voting on HR 2454, I demonstrate that only under the most permissive assumptions about constituency preferences could Obama have expected a successfully negotiated treaty to be ratified by the U.S. Senate. Given that ratification was virtually impossible—and telegraphed—there could be little expectation that China, Brazil, and many of the other countries would be willing to coordinate with the United States to achieve a mutually acceptable agreement. In effect, self-interest at the level of the individual drove opposition at the congressional level, which, in turn, constrained the successful negotiation of a treaty at the international level.

In terms of size, the U.S. Congress represents a much smaller group than the U.S. population as a whole and has the ability to pass legislation to impose constraints on the CO_2 emissions generated by that larger group. More influential representatives who have a stronger interest in the outcome can "pay" a greater proportion of the costs for achieving legislation and can act as the coordinator or punisher (DeSombre, 2007; Sandler, 2004), and by Olson's terms, this should provide for at least a modicum of support for an agreement to limit CO_2 production. That is, this smaller group—congressional legislators—has the ability to provide for at least some public good if political will to do so exists. Olson (1965) points out that the degree to which a collective outcome can be reached will be a function of the share, F_i, of the good that accrues to the largest member of the group (p. 28). One member of Congress with an abiding interest in the outcome can sponsor the legislation, push members toward support, and have his or her name attached to the resulting bill in return for compromises that shape that result. But each member of Congress represents a constituency that is both larger and less willing to pay the private cost of a public good. The willingness to pay the costs will be a function of how those costs are distributed across congressional districts. A representative from a state or district that would bear little cost from the regulation of CO_2 emissions but accrue significant benefits from climate stability would be willing to pay a higher political price than a senator or representative from a state with a carbon-based industrial infrastructure. Domestic politics might set up some horse trading, where the representative from the carbon-based state generates resources in return for support for, say, a climate change treaty. With a 19-vote majority in the Senate and a 75-vote majority in the House, the president could conceivably orchestrate the support required to ratify a climate treaty. The question is whether self-interested behavior at the level of the individual and congressional district trumps the public good at the national or international level.

Time influences support through reelection prospects, particularly when the costs of advocating carbon emission–reducing policy are high relative to future gains. The preferences of a congressional constituency is a function of the discounted future payoff ($p < 1$) and the level of uncertainty between cause and effect ($q < 1$). Congressional districts in which both p and q are low would be those in which the economic costs are borne disproportionately, such as ones with auto, coal, and oil industries. One difficulty in securing a successful policy is the need to make concessions to the representatives who would pay the constituent costs. The compromises made in order to ensure passage weaken legislative restrictions. We saw this repeatedly with CAFE standards and the automobile industry, carbon trading systems, and the coal and oil industries, as well as industries involved in power generation (Motavalli, 2011; UCS, 2003).

A different set of group dynamics operate at the international level, ones that Sandler (2004) and Olson (1965) suggest should have a better chance of achieving at least some of the public good. Copenhagen, 2009, however, failed to achieve a set of standards that even the U.S. Congress might have thought acceptable. The number of participants at the Copenhagen meetings was 114, accounting for roughly 80 percent of the national-level production of CO_2 emissions, and two of those countries, China and the United States, held oligopolistic interests. According to Olson, these countries should be able to form a coalition capable of securing a treaty because they have the ability to pay a greater share of costs and recoup a greater fraction of benefits. They also have the ability and interest in paying selective benefits and punishing defectors (DeSombre, 2007, 5–6). If one member is large and committed enough to absorb all the costs or in an oligopoly-sized group "where two or more members must act simultaneously, . . . there must be at least tacit coordination and cooperation" (Olson, 1965, p. 46). In fact, the Copenhagen meeting failed to achieve a commitment that approached even the level agreed to by U.S. congressional legislation when it should have been easy to do.

The implications of group-level interactions can be attributed to negotiations that transpire at two levels (Putnam, 1988). Even though structures are in place to arrive at a more optimal outcome at the level of the smaller group, interactions between group levels can constrain negotiating degrees of freedom, leaving the potential role of coordinator somewhat suspect, incapable of either making a credible commitment to absorb a significant share of the costs or credibly committing to punish violators in trying to define win sets (Putnam, 1988). In short, the ability of the smaller group at the international level is complicated by the requirement for obtaining political support at the congressional level. That, in turn, is complicated by the challenge of securing a commitment from citizens of the states required for ratification. Olson's model is not adequate when coordination must take place on multiple levels.

In an effort to coordinate a socially optimal outcome by committing to compliance, a two-level negotiating process is constrained by prospects for defection at the lowest level of aggregation, what might be best thought of as national level dynamics. Defection comes in two forms, voluntary and involuntary. An involuntary defection results from an inability to find a win-set at the international level that overlaps with the win-set at the national level, such that the failure to ratify a signed agreement leaves the agreement unenforceable (Putnam, 1988). The reasons for nonoverlapping win-sets between negotiated outcomes at the national and international level, particularly with regard to climate change legislation, could be tied to the discount rate (p) associated with expected benefits and uncertainty (q) surrounding the extent to which legislated adaptations would result in expected benefits. This would lead to an expectation that as the discount rate or level of uncertainty over cause and effect at the national level increases, the likelihood of reaching an enforceable international agreement declines.

Two concerns need to be addressed before moving toward testing expectations. First, considerable scholarship describes conditions for ratification of international environmental treaties (e.g., von Stein, 2008; Sandler, 2004; DeSombre, 2005 and 2007; Simmons and Hopkins, 2005). Von Stein, for example, provides evidence to suggest that flexibility between hard-law and soft-law provisions increases the prospects for ratification. Sandler and DeSombre, alternatively, describe the role of science, education, and negotiation in the ratification process. To get to the point of ratification, however, a treaty that can be submitted to the ratifying body for consideration must first exist. My argument moves this a step further back in the process to understand why some treaties get signed and submitted for ratification and others do not.

The second concern is that China is a critical player in the coordination necessary to forge an international climate treaty (Gray, 2009). Understanding preferences and political dynamics in China would be instrumental in understanding the outcome of the Copenhagen summit. My argument suggests otherwise. The domestic political environment in just one of the necessary oligopolistic players can be enough to prevent international cooperation. Put differently, a successful buy-in by one player provides a necessary condition for the coordination required to negotiate an international treaty. In that sense, one can be agnostic to the preferences of China and still understand the inability to negotiate a climate change treaty in Copenhagen, and even if both leaders hold preferences for a treaty, the expected defection in the ratification process can prevent coordination.

Evidence for the Expected Failure at Copenhagen

A U.S. vote on climate change regulation (HR 2454) and the president's trip to the Copenhagen conference are illustrations of this process. At the national level, there was a minimal level of support for climate change policy, and that did little to assuage the concerns of the international community. The 2009 G8 summit set target reductions in CO_2 emissions at 50 percent of 2005 levels by 2050 as a necessary step to prevent catastrophic climate changes; the Copenhagen summit was expected to generate an agreement that codified this level of reduction, which was much more stringent than the 18 percent the U.S. Congress was willing to allow (World Wildlife Fund, 2009). An agreement somewhere between the 18 percent agreed to by the U.S. House of Representatives and the 50 percent advocated by the G8 and climate change experts seemed a logical expectation. But, the Copenhagen meeting generated no agreement, confounding expectations about the effect of group size on public good provisions. To show why this might have happened, I use data on votes at the House level to draw inference about the likelihood of Senate ratification of a successfully negotiated treaty at Copenhagen.

Estimating directly the value of the discount rate and uncertainty, q and p respectively, is difficult because their true values are unknown. However, these values are linked in their political implications such that the value of p can be thought of as being a function of reelection cycles and q being a manipulable value that is also based on electoral exigencies. To the extent that the perceived costs to constituents from climate change regulation exceed the short-term benefits, any one elected official will hold a large discount rate and have an incentive to accentuate the uncertainties around anthropogenic causes of climate change. The incentives to doubt and to discount should be a function of short-term costs associated with jobs or income. To proxy the values of p and q, I use data on employment in three carbon-producing industries by congressional district: oil production, coal mining, and auto manufacturing. Data were derived from the 2008 County Business Patterns dataset in American Fact Finder from the U.S. Census Bureau, using North American Industry Classification System (NAICS) codes 2111, 21121, and 3361, respectively. The data on employment are recorded at the county level, which deviates somewhat from congressional district boundaries and were translated into congressional district data using Federal Information Processing Standard (FIPS) codes to allocate between county and congressional boundaries. When one county overlapped two congressional districts, employment data in the designated industries were counted twice, effectively assigned once for each district. This double counting biases the breadth of possible employment in the specific industries, but as a proxy of the political costs of climate change regulation, the attribution of one counties' labor force to two elected officials reflects more closely extant political concerns (Gartzke and Wrighton, 1998).

The data on county-industry employment are recorded by the Census Bureau in one of two ways: the actual number of employees or categorical groupings that provide a maximum and minimum within each range. According to the Census Bureau, proprietary information about actual employment can be withheld under U.S. Code Title 13 designed to protect data about individual businesses.[1]

The outcome variable is the vote on HR 2454, the American Clean Energy and Security Act of 2009. The final vote count was 219 in favor and 212 opposed to the legislation with three abstentions, a relatively close vote given that the distribution by party was 257 Democrats to 178 Republicans and the new president won by a significant margin. Roughly 18 percent of the Democratic caucus voted against the legislation and 5 percent of the Republican caucus voted for the bill. On the Republican side, the vote was primarily along party lines, and while this was largely followed by the Democratic side, there was significant deviation from the party position, suggesting that factors other than party affiliation were driving a significant proportion of the votes. In the aggregate, about 1 in 8 of the votes cast were across party lines (12 percent).

To test whether the individual congressperson's vote on HR 2454 was driven by demands that might point toward discounting future benefits in light of contemporary costs, I develop a simple logistic model that regresses votes for the HR 2454 on employment in carbon-producing industries. The results are presented in Table 2.1. This model is designed to give purchase on the likelihood of Senate ratification if the Copenhagen summit produced a treaty, in effect trying to tease out an answer to the counterfactual question.

1. Chuck Brady (Economic Unit of the U.S. Census Bureau), in discussion with the author, 2009.

Table 2.1 Carbon-Based Employment and Votes for Climate Change Legislation, HR 2454

Equation	*Variables*	*(1)* *HR 2454*	*(2)* *HR 2454*
HR 2454	Auto > 1,000 jobs		−0.981**
			(0.355)
	Oil > 1,000 jobs		−1.199*
			(0.500)
	Coal > 1,000 jobs		−0.264
			(0.522)
	party	−4.54**	−4.845**
		(.37)	(0.414)
	# jobs auto	−0.17**	
		(0.057)	
	# jobs oil	−0.75**	
		(0.20)	
	# jobs coal	−0.005	
		(0.13)	
	Constant	0.228**	2.191**
		(0.235)	(0.229)
	Observations	447	447

Robust standard errors in parentheses
* $p < 0.05$, ** $p < 0.01$

The results in Table 2.1 point to a significant relationship between industry interests in climate change legislation and votes by members of the U.S. House of Representatives. It should be clear from these results that votes were driven largely by party affiliation, but at the margins, the political cost of supporting this legislation was also instrumental. I use two indicators of the labor force tied to these three industries: the maximum number of employees per industry and a dichotomous indicator of more than 1,000 workers in each industry. I use a statistical estimator (logit) on the dichotomous indicator of whether a representative voted for HR 2454. A negatively signed coefficient suggests that the higher the value on employment in a carbon-based industry the less likely that a vote was in favor of the legislation. The coefficients, which can be difficult to interpret directly, can be presented as a measure of the probability of observing a positive vote on HR 2454. Table 2.2 presents these results.

Table 2.2 Probability of Voting Yes for HR 2454 (Percent)

Baseline	*Probability of Yes Vote*	*Change from Base*
Democratic District, Low Auto, Oil, and Coal jobs	90	
To:		
Baseline + > 1,000 Auto jobs	75	–15
Baseline + > 1,000 Auto and Oil jobs	39	–51

If I take as a baseline example a congressional district held by a Democrat with employment in each of the three industries falling below the 1,000 mark, there is a 90 percent chance that he or she will vote in favor or HR 2454. If that same Democratic district had employment in the auto industry that accounted for at least 1,000 jobs, the likelihood that the representative voted in favor of HR 2454 drops by 15 percent, to 75 percent. That is the effect of one industry with a significant number of jobs tied to it. If that Democratic district had accounted for over 1,000 jobs in both the automobile industry and oil production, refinement or extraction, the chances of that representative voting in favor of the climate change regulation bill falls by 51 percent over the baseline, meaning that the Democratic representative has only a 39 percent likelihood of voting for HR 2454. This seemingly restrictive condition of a Democrat from a district with significant employment in both the auto and oil industries is, in fact, true for 41 congressional districts spread across 25 states. And if 60 percent of these senators are going to vote against an international treaty, any ratification process is in jeopardy.

The results point in the same general direction—that employment, which serves as a proxy for short-term costs, decreases the chance of a particular congressperson voting for legislation to reduce CO_2 emissions. The model is simple; the results are generally strong. If we infer from these results that a high discount on future benefits reduces the ability of a group to provide for a collective good, then this provides some purchase on President Obama's apparent lack of interest in forging an international agreement at Copenhagen.

In the two-level framework articulated by Putnam (1988), President Obama's willingness to push for a strong agreement in Copenhagen should be a function of his ability to ratify any agreement generated from the negotiations process in Copenhagen. My argument is that the former could not happen because the latter was not going to happen, and the outcome was evident in the vote on HR 2454. If we assume that President Obama had a sincere preference for climate change regulation that was "better" than HR 2454, then absent any other constraint, he should have been able to coordinate with China to produce a treaty. Obama faced the prospects of what Putnam called involuntary defection at level two, and given this potential, it might be possible for the Chinese to recognize the difficulty faced by Obama and therefore refuse to coordinate on an agreement, even if it was in their interest to do so. Oddly enough, even if China, the United States, and every other country held a preference for a treaty that provided significant reductions in greenhouse gas emissions, the political process could constrain the options available.

According to a two-level analysis, even the small-group dynamics, under which a suboptimal but socially optimal outcome should be achievable, are not enough to overcome the potential for involuntary defection during the ratification process. Ratifying a treaty in the U.S. Senate requires 67 votes from senators subject to the same concerns about discounting future benefits relative to immediate costs that members of the House have to consider. So while there is no direct test of prospects for ratification had President Obama achieved an agreement in Copenhagen, the votes in the House for HR 2454 give us some understanding of how likely ratification might have been, assuming that the treaty out of Copenhagen was on terms consistent with HR 2454. The vote was close, particularly in a strongly Democratically controlled House, so the question Obama—and China, Brazil, and other instrumental countries—would have had to ask is whether the close vote in the House masked an ability of the president to generate ratification in the Senate, or whether it presaged the outcome of any such ratification vote. To test this, I take the votes for HR 2454 and aggregate them by state, making the assumption that votes in the House represent, at least in broad strokes, preferences of senators at the state level (Truman, 1956, p. 1024). Employment and, therefore, costs are state as well as congressional district concerns, and we might expect uncertainty, q, and the discount rate, p, to be high if employment in specific industries is high. Table 2.3 describes what might be thought of as the net expected votes in the Senate for ratification based on the House votes on HR 2454.

To generate these results, I made different assumptions about the net yea and nay district-level votes per state and then aggregated them to draw inference about possible Senate ratification. I started with the assumption that if the congressional delegation from a state voted, on average, against climate change legislation, senators in that state would also vote against a treaty that President Obama might sign. Because the vote on HR 2454 preceded the Copenhagen meeting by roughly six months, this information would have been available to both the U.S. and Chinese administrations and the world community. Given this assumption, the projected vote in the Senate would favor nonratification by 56 to 44 votes. This most simplistic form of aggregation assumes that both senators from a state would cast identical votes based on the majority position in the congressional districts, unlikely in practice, but an assumption that provides a good starting point.

A second approach is to assume that 50 percent of the net nay vote by state district translated into a yea vote by a senator that would provide President Obama with the slightest margin for ratification. That is, if the Senate delegation from each state split its votes when the aggregation of the House districts were against HR 2454—allotting 50 percent more votes for ratification than the congressional votes—there would be a narrow margin (+4 votes) for ratification of a treaty that was based on HR 2454 conditions. If President Obama could make that assumption, he would have gone to Copenhagen with a level of confidence that allowed him to coordinate an agreement under the conditions outlined by Olson (1965) and Sandler (2004). But this, too, is a broad assumption. The two most generous assumptions point to either a nearly certain chance of nonratification or the slimmest of margins

Table 2.3 Potential Senate Votes for Ratification of Climate Treaty

Assumptions about Vote Distribution	*Yea Votes*	*Nay Votes*	*Net Deficit for Ratification*
1) State votes assuming homogeneity with net district-level votes	44	56	–22
2) Assuming 50 percent nay votes are yea and 100 percent yea remain yea	72	28	+4
3) Votes if follow unanimous state delegation distribution (21 states)	18	24	
4) If split states follow district voting patterns by majority vote (added to unanimous states)	28	30	–20
If split delegations vote by following pattern (added to unanimous states):			
5) > = 70 percent favoring or opposing HR 2454	8	10	
6) District-level vote by a margin of +/- 2 votes	14	16	–26*

*This is a result of summing the columns across rows 3, 5, and 6 in favor of ratification of a treaty to reduce emissions by 18 percent, a level of compliance 32 percent below expectations for Copenhagen. In an attempt to gauge the likelihood that President Obama and the Chinese premier could have anticipated ratification of a more stringent treaty, given that they could coordinate and achieve a socially optimal agreement, I go back to the distribution of votes on HR 2454.

To consider a more realistic potential for ratification, I examine alternative distributions of votes by senator rather than by state. Absent a vote on ratification, this is a counterfactual exercise, but it helps to understand the absence of a treaty in Copenhagen. At the most basic level of examining the aggregation of state-level district votes, unanimity would be one indicator of whether or not both senators would vote for or against ratification if President Obama came back from Copenhagen with a treaty. Twenty-one congressional delegations voted unanimously on HR 2454, and 12 of those delegations voted against the bill. Starting at this point, the Senate votes would be 24 to 18 against ratification. Inferring the remainder of Senate votes by assuming that split delegations would vote based on the majority view of the district-level delegation generates a vote of 46–54 against, leaving the president short 20 votes—not great odds given the political climate and the discount rate for future benefits from regulation. A closer look at the distribution of those votes leaves little room for expectation of a positive ratification vote.

If the discount rate and uncertainty about cause and effect are important, employment should influence Senate votes just as it did the House votes, so I push the possible conditions for Senate ratification further. In districts where the sum of employment in the three carbon-producing industries is less than 15,000, representing the 25th percentile of employment in these industries, House votes in favor of HR 2454 were evenly split for and against the legislation (12 of 23 favoring the bill). But in districts where employment was above this 25th percentile, votes were against the bill by a margin of 63 percent to 37 percent (17 of 27 opposing the bill). Of the five states with greater than 15,000 employees in carbon-based industries, 80 percent voted unanimously against HR 2454. This represents about 25 percent of the votes needed to deny ratification.

Another way to evaluate the likelihood that the Senate would ratify a climate change treaty approximating the HR 2454 legislation is to look at the distribution of unanimous votes across congressional districts within states with high employment in carbon-producing industries. If the state delegation voted unanimously for or against HR 2454, it is reasonable to assume that the Senate delegation would too. Twenty-one state delegations voted unanimously on HR 2454. State-level unanimity was roughly 50 percent for and 50 percent against HR 2454 when employment levels in carbon-producing industries were in the bottom 25th percentile, but 80 percent opposed when employment was above the 25th percentile.[2] When short-term costs of jobs were at stake, it appears that the distribution of votes at district-level worked against support for HR 2454. When extrapolating from House votes to Senate votes, the unanimous state delegations are a minority (21 of 50), so it might be reasonable to expect that the 66 required votes could come from senators within 29 states splitting over the merits of a climate accord at congressional district level. That is, the assumption that the net vote count at state level reflects preferences of both senators is difficult to sustain.

The net vote count at congressional level includes those state delegations that were closely divided as well as those that were unanimous. In order to find the potential for those 20 votes that President Obama would need to squeeze out ratification of an international treaty, I disaggregate the close state congressional delegation into closely split and lopsided delegations. To generate these categories, I use two cutoffs for measuring the vote margin at state delegation level: a two-vote margin in favor or opposed to HR 2454 represents a closely split delegation, and a 70 percent or greater margin represents a lopsided congressional delegation. For example, if the votes in a state were five to four in favor (or opposed), this might suggest that at the ratification stage the two senators from that state might also split their vote. Conversely, if the margin in favor (or opposed) was

2. There were five states in which the delegation was unanimous and employment was greater than 15,000, with four voting against HR 2454. Twenty-five percent of the votes needed to prevent ratification were cast by these four states.

greater than 70 percent, it might be more likely that the distribution of Senate votes from that state will be unified. In states where the House delegation did not vote unanimously on HR 2454, nine of them had vote counts that were by margins of greater than 70 percent for or against the bill, five of which voted against HR 2454. When the state delegation vote was close in the House, delegates rejected HR 2454 eight of 15 times. Assuming unanimous state delegations voted homogeneously, and overwhelming or closely split delegations vote as described above, the expected vote for ratification of a treaty based on HR 2454 principles would be 40 in favor, 60 against. Ratification would fall short by 27 votes.

Given the hypothetical distribution of Senate votes based on alternative scenarios of aggregating known district-level votes in the House, it is very hard to see how President Obama, China, or anybody else could see the potential for ratification of a climate treaty based on the outlines of HR 2454. And, importantly, the legislation encapsulated in HR 2454 required 32 percent fewer restrictions on CO_2 emissions than the target set by the countries participating in Copenhagen. So even if there were a window for ratification, the potential treaty might have either been too tepid for China and the rest of the world community, or far too restrictive for the U.S. Senate to take it seriously. There is only one set of conditions where the votes appear to support ratification, and this outcome is 1) a close vote and 2) based on the most simplifying and difficult to justify assumption about potential voting behavior. Things did not look good for the ratification of a treaty negotiated at Copenhagen and signed by President Obama.

Implications

The analysis of the possible distribution of Senate ratification votes is counterfactual, but there are fairly dramatic implications that derive from the analysis. I highlight these in terms of theoretical and policy suggestions, starting with policy. Some of the main points are that 1) there was never the opportunity for the Senate to consider an international climate treaty negotiated at the Copenhagen summit,

2) the employment in carbon-based industries helps to account for votes at the congressional level, and 3) that under a series of plausible but hypothetical conditions, it is very difficult to find the right mix of conditions for a possible Senate ratification of even a tepid treaty coming out of Copenhagen. The first of these points should be puzzling: Why wouldn't the United States push hard for a treaty to sign and ratify? The second reveals the political costs associated with climate change policies and the way parochial interests shape how benefits, costs, and expectations are managed in the political arena. The third highlights the multiple levels of political interactions that permeate something as complex as a climate change treaty. It is not enough to think about this in terms of international organizations or institutions, nor is it sufficient to think in terms of lobbying or other forms of political participation. Individual-level behavior, national-level politics, and international treaties are all tied up in the same process. That is what makes this problem a hard one to address.

To devise a possible resolution, we have to keep in mind the core components of the collective action problem: costs (C), shares that accrue to individuals and groups (F), benefits (V), uncertainty (p), and the future discount (q). Shares are a function of group size and, for the most part, are not easily manipulable. Benefits include a stable planet and are worth fighting for. The costs, on the other hand, might come in two forms. One, a planet that is virtually uninhabitable in the way it is now; the other is the individual contribution to stem pressures on our climate systems. These latter costs are the ones that you and I pay, in the form of the expense for low-energy lightbulbs, cars that produce low levels of CO_2 gases, a new norm for the comfort level in residences, a DVR that does not start up instantly. Some of these things will cost each of us money out of our pockets. As demonstrated earlier, our willingness to pay these costs in pursuit of the benefits will be a function of how certain we are of the causal process and how long into the future we consider the benefits to be worth the costs. Both of these can be manipulated. To achieve the public good of climate stability, it is advantageous to manipulate these factors to increase the level of participation. We also have to keep in mind that these factors are influential at all three levels at which actions can impact the outcome. But there is also an assumption that if compliance was easy at the level of the individual, it would be easy at the congressional level as well, which, in turn, would make international leaders more likely to negotiate and sign a climate change treaty. The reverse order of influencing the outcome can also work, and in some political systems, top-down action may be the only way. But in most countries of the world today, and certainly in the United States, compliance—either coerced or willingly—at the individual level has the potential to shape the process at the higher levels of group aggregation and vice versa.

Sandler (2004) compares the results from international efforts to control ozone-producing emissions with those that generate climate change. He argues that, when costs figure more prominently than benefits in the policy discussions, the prospects for successful regulation decline. If he is right, how we generate our understanding of these relative costs is critical. In climate change issues, costs are borne immediately, while benefits accrue only in the future, allowing individuals to discount their value. My evidence suggests that job-related costs outweigh the influence of climate reduction benefits in the willingness of people to support a collective good policy. Until the discount rate for benefits from climate-stabilizing policies decreases relative to current costs, any international treaty will be subject to the constraints of a two-level process where the efficiencies of small-group dynamics will succumb to the exigencies of larger-group demands. Education campaigns that focus more closely on benefits rather than costs could change this, but there are competing interests with a say in such campaigns. Industrial interests, for example, have incentives to emphasize costs and uncertainty, because they would have to retool and determine how to market goods that compete with similar, cheaper products that are not green. We should keep in mind, however, that one of the supporters of HR 2454 was Ford Motor Company, an industrial concern with many of the same short-term interests of other auto companies, as well as the coal or oil industries. Ford has a long tradition of supporting innovative and green technology and is an example of an industrial concern participating in the pursuit of social outputs. Even industries that are conscious of short-term costs can emphasize the benefits from future changes through marketing. Wal-Mart, for example, shifted from a mix of incandescent and florescent to marketing only low-energy florescent bulbs. It might have been a narrow business decision motivated in part by legislative restrictions (2007), but it is marketed as a sound environmental initiative.

From a political vantage point, my evidence suggests that supporters of climate change regulation in Congress might have been misguided in bringing legislation to the floor that would reveal margins of support *before* an international treaty could be negotiated. Part of the result at Copenhagen derived from the lack of coordination among the top emitting countries, potentially driven by other countries' anticipation of the failure of ratification in the U.S. Senate. A simple analysis of the votes on legislation similar to, but even weaker than, Copenhagen was sufficient to show that President Obama had a weak hand going into the summit. If the political process were orchestrated with more foresight, it may have succeeded in overcoming the liabilities from revealed preferences and constrained outcomes. Understanding the implications of an involuntary defection in the ratification process could have changed the dynamics in such a way as to put the weight of resistance on the opponents, as Putnam (1988) describes in negotiations over the Bonn Accord.

The costs of taking steps to reduce our consumption are not technically fixed in time. But for the most part, energy efficient alternatives tend to be more expensive than the less-efficient options. Spiral florescent lightbulbs cost $15–20 in 1997, and today cost $2–4. An incandescent bulb, though, costs less than a dollar. Though incandescent bulbs did become relatively more expensive as the market for LCD bulbs increased, the more efficient technology costs more. To rationalize the extra outlay, potential consumers would have to be convinced that a) it will have an impact, and b) that their payoff will be in the reasonably near future. This is where the political competition comes into play.

There are also implications for how we think about some types of collective goods. From a theoretical perspective, for example, our understanding of the provision of some collective goods requires us to go beyond models that account for group size, coordination among oligopolies, or costs and benefits to consider how politics affect the efficiencies of groups. Since public goods reflect political choices and often derive support across different constituencies, models of optimal performance must consider the interaction among groups of various sizes and constituencies, and how perceptions of the prospects for acquiring a good commensurate with its costs can be manipulated. In Olson's theory of group efficiency, he relied on the effectiveness of firms in the market. Many public goods reflect externalities to the market and rely on public regulation and coordination, or self-regulation (Ostrom, 1999), and therefore private incentives to prevent coordination.

It seems from the outcome of the U.S. domestic effort to secure climate change regulations and the president's effort to codify these efforts among international actors that too much information at one group level can influence the outcome of another group at another level. Groups are not independent when some public good issues are at stake, and the ability of groups to secure some provision of that good might be hamstrung by the efforts of another group, even if that other group was able to achieve some level of social optimality. In this sense, the ability to coordinate methods to achieve a collective good is complicated by telegraphing the likely outcome of a ratification vote. If the information about ratification in the U.S. Senate had been kept private, President Obama might have been able to negotiate a treaty by coordinating with the Chinese and pressuring the Senate to ratify after the fact. But by making the likelihood of ratification public through votes on HR 2454, others may have had little interest in coordinating on a treaty that could not succeed.

Although an international climate change treaty might be unique in how uncertainty, the window of the future, and collective action at multiple levels interact, international agreements clearly are not. The ability to coordinate among a small group of actors on the international scene remains relevant, but what we have learned about the political process might point us toward more efficient or effective strategies to coordinate and ratify an agreement among international actors when confronted with complex conditions for treaty obligations, multilevel interactions, and critical social policy.

One last note about the process at the Copenhagen climate summit is in order. Although President Obama came home without a treaty, and although the House vote on HR 2454 was successful by a slim margin, the United States has put in place a number of steps consistent with the objectives of the Copenhagen initiative. The most prominent of these may be the new CAFE standards mandating 54 miles per gallon by 2025. Not quite up to European standards and on par with Chinese ones, these new standards double the current ones mandated by the government. Wind and solar energy companies have taken substantial steps to increase production, spurred by incentives as well as mandated quotas. 2009 was the first year in decades that the U.S. contribution to CO_2 gases has declined from year-on-year levels, thanks in part to a deep recession and high gas prices. Smaller incremental steps are possible at individual and national levels, even though without an international treaty the collective good will not be achieved in any reasonable time frame. There have been successes at the international level, and understanding the reasons why some global commons issues are tractable to international agreements could be used to shape the processes with regard to climate change.

But neither was Copenhagen the last opportunity to generate an international treaty. The series of climate meetings in Doha, 2012, Warsaw, 2013, and New York, 2014, pave the way toward the Paris summit in 2015. The expectations at Paris are that the countries of the world will do what they could not at Copenhagen: agree to a set of constraints that will substantially reduce the amount of CO_2 emitted into the atmosphere. Paris holds out promise—in part because of the increasing pressures on our climate systems, in part because of increased global attention, and in part because of the lessons learned at Copenhagen. And, even though Paris is not certain to bring about an enforceable agreement, six years out from Copenhagen, there may be a sense that the ability to discount future benefits is narrowing and with the more recent IPPC report of 2014, an understanding of the anthropogenic drivers of climate change is more certain.

The climate meetings in New York in September of 2014 and the protests there and around the world demonstrated that there is public concern about climate change and support for a restrictive treaty. But a series of protests show only one side of the debate. Those opposed to restrictions because of jobs or other perceived economic costs largely sat on the sidelines. They get their chance in the process of developing national positions and shaping legislation.

The host of the 2015 summit, France, holds expectations for an enforceable treaty committing nations to carbon reductions for the first time since the Kyoto agreement. And, according to the French Foreign Ministry, they expect to push for a 40 percent reduction in greenhouse gas emissions by 2030 and a 60 percent reduction from 1990 levels by 2040 (France Diplomatie, 2014). These ambitious goals are largely in line with the initial hopes for Copenhagen, though we will have lost nearly a decade in implementation.

3

♦ ♦ ♦

LESSONS TO LEARN

OZONE AND CO_2

Collective goods impact our lives at local, national, and global levels. The ozone layer, as well as whales and wolves, are examples of such goods, and, like climate, ones that were negatively affected by human activity. In all three cases, we took actions that reversed some of the negative effects. Though none of these are on the same scale of magnitude as climate change, and though climate change is one of the largest and most serious of the challenges we face, these environmental issues are examples of successes achieved by the world community. I will examine these examples in terms of how we might manage CO_2 consumption associated with climate change.

Rather than account for the political processes or the outcomes of these efforts, my objectives here are to highlight the core components of the collective action problem and how overcoming it can have significant and measureable benefits, and failing to do so, significant costs. Dealing with each of these environmental issues came up against the same expected payoff dynamic as climate change does today. But in each of these examples, we proved to be more adaptable than we have been in our efforts to ward off the destruction that a rapidly changing climate would bring. None of these examples presented a challenge of the complexity of climate change, but each did require individuals, states, and the international community to overcome the incentives to continue harming the planet. The same core question of the future value—which would derive from making hard choices now to deal with climate change—were critical. Constituency groups for each of the issues existed. In spite of this, a cooperative arrangement was reached, and in all cases, agreements and policy changes were able to turn things around.

To go back to my initial framework, for individuals participating in a collective good, a net positive value is expected on returns on investments such that $E(X) = ((V_i * q_i * p_i) - C_i)$. The expected payoff from addressing the environmental problem is a function of the costs (C_i), the benefits of success (V_i), the uncertainty about cause and effect (q_i), and the degree to which future outcomes are discounted relative to cost paid in the present (p_i). When the payoff is less than the effort, there will be free riders. Technology accounts for some of our ability to respond to environmental pressures. But particularly with regard to selective incentives, politics can play a central role by decreasing costs and influencing uncertainty about cause and effect, or the future discount rate. This is simple in theory but not in practice.

Factors work together to make free riding a rational choice, thereby making collective action difficult. The size of groups is critical, and smaller groups are more likely to achieve a better level of success relative to larger groups. When we think about this from the perspective of a political process, uncertainty and future discount rate—p and q—explain many of the outcomes. The examples I discuss highlight the role of discounting future benefits and uncertainty, and the political processes or hurdles that must be managed to address climate change. However, although there are consequences to eliminating the top predator in the American ecosystem, the extinction of wolves does not carry the same catastrophic consequences of overheating the planet. If we eliminate the top predator in the U.S. Great Lakes region and the West from the ecosystem, we will suffer downstream consequences but not on the same order of magnitude that we will from dramatic weather events, shifting agriculture patterns, melting of ice shelves and glaciers, and the loss of coastal habitation.

But the processes of saving wolves, whales, and the ozone layer involved overcoming similar hurdles. We can learn from these comparatively smaller but successful efforts how to address the really big one. In this chapter, I start with the problem of depleting the ozone layer and our collective efforts to revive it.

Ozone Depletion and the Montreal Protocol

On December 21, 1987, President Ronald Reagan transmitted the Montreal Protocol on Substances that Deplete the Ozone Layer to the Senate for advice and consent. In March 1988, the treaty was ratified unanimously by the U.S. Senate on a vote of 83–0, even though the Senate was split between 53 Republicans and 47 Democrats. The Montreal agreement and its amendments have been heralded as the signature effort by the world community to address anthropogenic causes of environmental challenges (Speth, 2004). From the start of formal negotiations to limit ozone depleting gases to ratification by the U.S. Senate, the process took about a year. By contrast, the Kyoto Protocol on climate change was adopted in 1997; the president signed the treaty a year later, but it was never ratified by the U.S. Senate. It technically came into force in 2005, and Canada withdrew in 2011. Kyoto simply couldn't save us from ourselves. Something different transpired to get the world community on board with restricting ozone gases that is not in play when it comes to greenhouse gases. The explanation could be in the political processes, the technological challenges, or differences in the way the two issues are addressed.

It would not serve me well to provide a technical description for how certain human-made gases deplete the ozone layer that surrounds the earth, primarily in the stratosphere that is roughly 15 miles above the earth's surface. I leave those important technical details for the chemist. What I will do is describe the process sufficiently to develop some of the differences between ozone hole depletion and CO_2 concentrations. Some of our understanding of the differences in resolving one problem while struggling with the other can be found in the chemical processes themselves.

The core of the ozone depletion process involves the emission of carbon-based gases from industrial and consumer consumption into the upper atmosphere; these gases break down into forms of chlorine molecules that then interact with ozone. Ozone is a molecule with three oxygen atoms, rather than the more normal two, and in this state is a more fragile molecule. The ozone layer is made up of these molecules. It is a thin layer of our upper atmosphere, but an important one, that shields humans from the intense power of the sun. Chlorine interacts with the ozone molecules and breaks them apart. The more chlorine there is in the stratosphere, the more extensive and faster the breakdown of the ozone layer. This interaction was hypothesized in the mid-1970s, but there was little firm evidence until the mid-1980s and beyond (Benedict, 1991).

There are six main ozone-depleting gases that fall under the general names of chlorofluorocarbons (CFCs) and halogens. One of the unique things about these gases is that they are all chemicals manufactured in factories. Not one is a naturally occurring element that you would find on the periodic table. As a result, the way nature interacts with these manufactured chemicals caused part of our natural ecosystem to begin to break down. The reason we made such a disastrous set of chemicals in the first place is logical: they were industrially useful, and the downstream consequences were unknown.

CFCs and other ozone-depleting chemicals have industrial properties superior to the chemicals they replaced, except that they have catastrophic consequences for that part of our environment that few understood at the time. These CFCs, it turns out, have household names that many of a particular generation would recognize. Freon, for example, was the proprietary name of the CFC 13 gas developed by DuPont, once used in car air-conditioning systems and refrigerators. When an air conditioner was recharged or an old refrigerator discarded, Freon was inadvertently released into the stratosphere. A certified process is now required to recharge an air conditioner, and refrigerators must have their compressors removed before they can be discarded. These relatively new restrictions are designed to prevent the escape of CFC gases, thereby reducing the depletion of ozone. The restrictions on consumption derive directly from international agreements that form the Ozone Treaty, which imposes national-level restrictions on production of CFC gases. State governments, in turn, ensure compliance by imposing constraints on the local use of ozone-depleting gases. Deodorant, household cleaning sprays, and almost any other fluid propelled out of a can once used an aerosol process based on CFCs. Today, we have pump spray bottles, and where gases are necessary, those used are less harmful to the ozone layer. Without the ozone layer, recreation, food production, health, and longevity were at risk, and the costs of these restrictions were absorbed willingly.

A large part of the resolution had to happen at the individual level of consumption. Yet most people today are not keenly aware that we have changed our behaviors and consumption patterns in ways that appear to be having a very positive effect on the ozone layer (Speth, 2004; Slaper et al., 1998). But we did recognize that we had to change to fend off the deleterious consequences of ozone depletion. Within a generation, old methods are obsolete, and the costs of those changes are unknown to many current consumers. Costs (C_i), then, became small relative to the value of the ozone layer (V_i), and we came to understand that the consequences were much more certain and much more immediate than initially thought (q_i and p_i).

Consequences of Ozone Depletion

When we go to the beach, a picnic, or the golf course, the application of sunscreen has become second nature. The sunscreen is to protect us from skin cancer that is often the result of overexposure to ultraviolet (UV) radiation from the sun. In the old days, a sunburn was just a consequence of sun exposure; today, it is the cause of worries about skin cancer. One critical role of ozone is to block the UV–B rays from reaching the surface of the earth and, therefore, our bodies. As the ozone layer gets thinner, more of the harmful rays of the sun reach us, and subsequently, the rate of skin cancer increases. These human costs from a depleted ozone layer have quantifiable social consequences that were not always well understood.

UV radiation comes in two forms, UV–A and UV–B, and it is generally the UV–B rays that cause the most damage. To put this into context, for every 1 percent drop in the thickness of our ozone layer, there is a corresponding 2 percent increase in the amount of UV–B rays reaching the earth. In 1988, it was estimated that a 2 percent drop in the ozone layer would result in an increase of about 145,000 new skin cancer cases by 2025, something on the order of an increase of 6–8 percent in skin cancer rates (Brunnee, 1988). Estimates by the National Oceanographic and Atmospheric Administration (NOAA) suggested that a significant decline in the amount of ozone in the upper atmosphere started in the 1975–1980 period, and that without any remedial action to reduce the trauma to the ozone layer, by the year 2000, there would be an annual increase in UV doses at the earth's surface of about 15 percent (Slaper et al., 1998). If a 2 percent drop in ozone would generate a 4 percent increase in UV–B rays, and a 2 percent increase in those rays generated an extra 145,000 cases of skin cancer, the consequences of a 15 percent annual increase in UV exposure could be catastrophic.

But skin cancer is not the only problem with ozone depletion. If UV–B rays can turn normal human skin cells into mutations that cause cancer, it must also afflict other cellular structures. Evidence demonstrated that UV–B can affect the growth and the forms that plant cells take, as well as altering the growth of marine phytoplankton. Phytoplankton are very small plants that help turn carbon dioxide into oxygen and are critical for a stable ecosystem. In short, people, crops, and fisheries were at risk from a dramatic increase in UV–B radiation (Speth, 2004). Experiments also showed that up to two-thirds of the plants tested had some form of negative consequence from UV rays, including consequences for yields, leaf structure, biological functioning, and germination (Brunnee, 1988; Benedick, 1991). If overexposure to UV radiation were to alter various aspects of the food chain, the potential for widespread disruption posed a significant problem for humans, and this would be a collective problem that we could not solve simply with sunscreen.

The first studies to draw possible linkages between CFC gases, ozone depletion, and skin cancers can be dated to the mid-1970s, and strong empirical evidence that CFCs were doing damage to the ozone layer was not available for another decade. What is referred to as the ozone hole was not recognized until 1985 (Speth, 2004). The ozone hole was a seasonal change in the depth of the ozone layer over the Antarctic; the hole was reasonably small in the winter but grew significantly during the summer months, eventually encompassing New Zealand and large parts of Australia. Skin cancer rates in those countries are some of the highest in the world.

Studies were beginning to make the theoretical connection between an increasingly prevalent man-made gas with numerous industrial and consumer applications and risks to human health and habitation. But at that time, evidence was limited and far from certain, and projections for the worst of the outcomes were pushed 25 to 50 years into the future (Benedict, 2005; Septh, 2004; Brunnee, 1988). Recognizing the problem with CFCs required trying to secure the collective good of a stable ozone layer when participation had to satisfy the conditions described by Olson (1965) and contend with the role of uncertainty and future discounting. Under many conditions, it would have been easy to free ride and politically astute to resist signing an international treaty, or fail to ratify a treaty that the president signed. But in reality, the time between President Reagan signing the Ozone Treaty and the Senate ratifying it was a matter of months. The time from the first studies suggesting CFCs might be destroying the ozone layer and the consequences of that to the negotiation of the Ozone Treaty was a decade. Since the signing of the treaty, there is evidence of a considerable reduction in ozone-depleting gases in the stratosphere, as well as significant regeneration of the ozone layer. Studies suggesting that CO_2 emissions and other GHGs could generate planetary warming with significant climatic consequences have a longer pedigree than those concerning ozone depletion, and evidence of cause and effect has accumulated more rapidly, but world leaders could not come to an agreement in Copenhagen in 2009.

The consequences of CO_2 emissions became known at roughly the same time as those of CFCs, but nearly 30 years after the successful Montreal Protocol that banned CFCs, generating an international agreement that will constrain our use of CO_2-producing gases remains a struggle. There are differences between the ozone and greenhouse gases debates, and understanding one may help understand how to develop a better strategy for addressing the other. The problem lies not in access to technology but in the politics, and my focus is on four differences that shaped the political debate: how CFCs were produced, who produces them, the role of science, and how the future benefits were viewed relative to the immediate costs.

The Production of CFCs

Securing the ozone layer is a collective good, and Olson's (1965) model points to costs, benefits, and relative shares of the good as critical factors in understanding the global community's response to CFCs and CO_2 emissions. We also have to add the role of uncertainty and the discounting of the future. These factors are affected by methods of production and costs of change, and politics can reside at the center of making necessary changes.

There is a considerable difference between how CFC and CO_2 gases are produced. CFCs are manufactured products. A company invests in research and development of a specific product because it has potential for industrial or consumer applications. CO_2 gases, for the most part, are natural, at least in the sense that carbon is an element in nature, found on the periodic table. All humans do with the element is move it around—for example, combine it with oxygen to create the molecule CO_2 in the combustion process. A gallon of gasoline weighs about seven pounds, and burning that gas generates about 20 pounds of CO_2. Every time we consume carbon-based products or processes, such as when flying a plane or driving a car, we generate CO_2—one of the primary greenhouse gases that is warming the planet.

The DuPont Corporation held patents on many of the most harmful CFC products, but nobody holds a patent on CO_2 or carbon. Alternatives were found to manufacturing CFCs. There is, however, no alternative to CO_2 as a by-product of the consumption of carbon-based fuels. The only way to reduce CO_2 generation is to reduce consumption, either by changing behavior or changing efficiencies. Industry found an alternative to CFCs that met certain standards for application and affordability, and industry, in theory, should have been more willing to adapt to meet the needs of the environment.

In the mid-1970s, when science was beginning to question the impact of CFC gases on ozone, DuPont accounted for nearly 50 percent of the total U.S. Freon production and upwards of 25 percent of world production (Molina and Rowland, 1974; Maxwell and Bricoe, 1997). There is no comparable company in the CO_2 debate. Big oil companies, coal, and the entire manufacturing industry are all significant contributors, and so are individual consumers. Today, consumers have choices in many of the products that release CO_2 into the atmosphere, unlike with CFCs. A refrigerator came in a choice of sizes, colors, and brands, but Freon was not a choice. Today, consumers can buy more or less efficient cars, heat their homes to higher or lower temperatures, and decide whether to drive, walk, or take a bus. As consumers, individuals played a smaller—and different—role in when, where, and how CFC gases were produced.

From this perspective, consumers contributed differently to the solution to the problem of ozone depletion. The budding environmental community pushed for a ban on CFCs, primarily in the aerosol form. Such a ban had implications on deodorant and hair spray, but an individual's share of the good was not a function of how much cost was paid or how efficient the consumption. DuPont, on the other hand, was, as Olson's model would suggest, the oligarch that could absorb the costs and share disproportionately in the benefits. And that they did.

According to a history of DuPont's strategy to address the potential ban on CFCs, DuPont initially resisted a call to restrict or ban Freon. Congress, under environmentalist pressure, held hearings on the new science of ozone depletion, and legislation was introduced to ban CFCs as propellants at both the national and state levels. By the mid-1970s, nearly a dozen states had taken up the issue of banning CFC aerosols (Maxwell and Bricoe, 1997). DuPont then began scientific studies and explored alternatives to the standard CFCs in use at the time. Having alternatives ready would allow them to continue to lead the market in event of a ban. Although the dominant player was not fully committed by the early 1980s, scientific evidence and market processes combined to help turn the tide in the CFC debate. By the late 1980s, there was so little resistance to regulating ozone-depleting gases that not only was the Montreal treaty overwhelmingly ratified by the Senate, but an addition to the Clean Air Act that severely restricted CFC gases passed both the House and Senate by large majorities (HR 3030; Senate 1630 of the 101st Congress). The Senate bill passed with only 11 dissenting votes and the House with only 21. These lopsided votes would be unlikely if 1) the science was not sound enough to reduce uncertainty, and 2) the public didn't overwhelmingly support passage. What we know from the climate change debate is that the sense of scientific uncertainty can be manipulated such that public support becomes too weak to get elected officials to cross lines. The interpretation of science matters.

The Role of Science

Even though the science behind greenhouse gases, global warming, and climate change has a longer pedigree than that of ozone depletion, our understanding of both from a practical perspective can be thought of as contemporaneous. British scientist John Tyndall (1873; Fleming, 1998) demonstrated in 1860 that CO_2 molecules absorbed heat. We, therefore, had a very early understanding of what fossil fuel consumption might do to the atmosphere. If CO_2 absorbs heat and we increase the amount of CO_2 in the atmosphere, we generate the greenhouse effect that contributes to an increase in global temperatures. Charles Keeling began the first systematic measurements of CO_2 concentrations in 1958. By 1960, he had data to show that observed increases coincided with increases in fossil fuel consumption. As evidence began to accumulate in the 1970s and 1980s, NASA scientist James Hansen coauthored an article in *Science* making a strong case for anthropogenic causes of global warming and made dire predictions for the future of the planet if consumption patterns did not change (Hansen et al., 1981). Hansen made his case before Congress, and he is considered to be one of the first to highlight a potential consequence of global warming and climate change. This period roughly coincides with the discovery that ozone-depleting gases were affecting UV radiation on Earth. The ozone dilemma was resolved relatively quickly, while the greenhouse problem generates considerable resistance. My interpretation of Olson's collective action problem should help shed light on why.

For Olson, each individual had to maximize their utility, or at least get a positive value relative to whatever costs they paid toward a solution. When each individual will recoup at least as much relative to the group pushing the changes and this payback provides a net positive gain, the individual will participate. In its simplest form $E(X) = (V_i - C_i)$ must be positive. If the value (V_i) is sufficiently high, then each individual, community, and country should be willing to pay a significant cost (C_i) to generate the benefit. If the costs of change are manageable, then a high-value outcome with low costs would be easy to achieve. Science should be a big part of determining costs and value of reducing ozone depletion. Science should also play a significant part in the level of uncertainty about cause and effect, as well as the understanding of how much time until the harm was irretrievable. It is possible that if the difference between costs and value of a thick ozone layer was sufficiently great then group size would not matter as much as Olson anticipated.

When it came to regulating CFC production, the president, senators, and congressional representatives had to have a sense of how costly a ban would be, both for them personally and for the economy. If DuPont was going to resist any efforts to ban CFCs strenuously, it would pose a problem for the president signing a treaty, the Senate ratifying it, and Congress enacting legislation that would implement the obligations under the treaty. In this case, all the pieces fell into place and the United States implemented a ban on the production of CFC gases. The votes were so overwhelming that there was virtually no political resistance.

When the science relating CFC gases to the depletion of the ozone layer first surfaced, DuPont resisted change, because it appeared to influence their business model. They were not the only producer of CFC gases, but they were by far the most significant one. DuPont scientists worked on two complementary tasks. First, they studied the effect of CFC gases on ozone depletion and the related question of the consequences of a depleted ozone layer. Second, they began the research and development required to generate alternatives to CFCs. In testimony before Congress, the CEO of DuPont committed to change if indeed evidence supported the bad consequence argument in the CFC/ozone debate. He went on record to say that his company would do the right thing (Maxwell and Briscoe, 1997).

A few things came together to make this an easy case for the world community, DuPont, and elected officials to do the right thing. About the time that the ozone hole was discovered over the Antarctic region, DuPont's scientists were confirming the results of other scientists that CFCs were indeed thinning the ozone layer and that there would be significant long-term consequences. At the same time, they were working toward an alternative to the standard CFC gases. The future discount rate for maintaining the status quo was very small, the benefits were clear, and the cost could be borne by an oligarchic actor. The one actor that was really able to mobilize political resistance had the potential for a new monopoly on the production of alternatives to the current industrial uses for CFCs. If a ban was enacted, DuPont was in position to capitalize by marketing the only viable alternative. With no need to prevaricate, the company became an advocate for a global ban. There is no similar actor for CO_2 gasses: there is no manufacturing process that creates CO_2 outside of consumption processes, and those consumers are ubiquitous—literally most of humanity—making it is so much easier for each individual to free ride.

Evidence in the Congressional Record

The president signed a treaty banning the production and use of ozone-depleting gases that was negotiated in Montreal, and the U.S. Senate ratified it overwhelmingly. Neither, however, provided for an action clause that would force the United States to meet its treaty obligations. Amendments to the existing Clean Air Act provided the mechanism for implementation. And even though the vote in the

House and Senate that created the legal framework for complying with the Ozone Treaty was overwhelmingly in favor of the legislation, it was not unanimous. Roughly 5 percent and 10 percent of each chamber, respectively, voted against the amendments to the Clean Air Act. According to my argument, the opposition to the legislation should be explainable by recourse to private interests.

Olson's argument about collective action would point to the expected value [E(v)] of participating at the individual level, so one way to think about this is to ask who is paying more than they would expect to get back. One of the main differences between the CO_2 and the ozone debates is that, in the case of ozone, the largest possible group—individual consumers—were not hugely affected by the outcome. Most individuals did not care how deodorant was propelled or how air conditioners and refrigerators worked, as long as they did work. If there was an alternative propellant that did less harm to the world, most individuals did not object to the switch, as long as the cost was not unbearably high. The one glaring exception was individuals whose jobs were tied to the continued manufacture of CFC chemicals, effectively a private interest that might drive participation in pursuit of a collective good.

If the collective action arguments are right, and people respond based on their individual expectations for a return relative to the cost of achieving the common good, then those who would pay the highest cost might be those whose incomes depend on the status quo. If individual constituents' jobs are at risk, their congressional representative would have an incentive to vote against government-imposed restrictions, even if the consequences of the status quo have a large environmental cost. And while each individual worker in a CFC plant might not leave an observable trace of his or her response, the workers' elected officials would. The data would show if those who voted against legislation to implement the Ozone Treaty came from districts with workers in the CFC industry. If the pattern was similar to the votes on HR 2454 (CO_2), the ozone debate would shed light on the contemporary politics behind climate change policy.

We know that the Senate ratified the international treaty unanimously, but we also know that 11 of the 100 senators and 21 of

435 representatives voted against amendments to the Clean Air Act that would give teeth to our treaty obligations. To test for this private incentive to vote against the tide of opinion, science, and history, I used data on carbon-based chemical manufacturing employment by congressional district to see if jobs in the CFC-producing industries predict votes in the House of Representatives. As with the data on CO_2-producing industries used in a previous chapter, I acquired data from the U.S. Census Bureau based on the Standard Industrial Classification (SIC) system that identified employment in the Industrial Organic Chemicals industry. Because specific data on production locations are not available, and there is no specific SIC coding for CFC chemicals, I have to accept that I am using a broad brush to paint a fine-lined picture. In effect, I am including employment in other industrial organic chemicals in jobs specific to the production of CFCs. These data are recorded at the county level. I overlaid them with the congressional district and aggregated employment data at the district level. If private incentives are at the core of participating to secure a collective good, the number of jobs tied to CFC production should predict a congressperson's vote on HR 3030.

Table 3.1 presents a basic model of the relationship between jobs in CFC industries and votes on HR 3030, the amendments to the Clean Air Act legislation that put in place some of the strictest CFC regulations to date. The way to interpret this model is to ask whether the data suggest that in a district with more jobs tied to CFC production, the chance of the congressperson from that district voting against HR 3030 is greater. The answer to that is yes. When viewed from the perspective of the likelihood of a vote against the legislation, the results point to jobs as a significant factor in an individual "no" vote. The outcome variable is coded 1 if the congressperson voted no, so a positively signed coefficient shows that as the number of related jobs goes up, the chances of observing the "no" vote does too. Moreover, if the senator for that state voted "no" to the Senate version of the bill (S 1630), there is a much greater chance that the congressperson did as well (also a positive coefficient). This result is not trivial in terms of the estimated probability of observing a congressperson voting "no."

Table 3.1 House Vote on HR 3030 101st Congress, Outcome Is Voting "No"

Max employees	0.0006**
	(0.0002)
# Establishments	−0.12*
	(0.052)
Senate no vote	1.6**
	(0.48)
Large employer	−2.9*
	(1.4)
Constant	−3.04**
	(0.34)

Observations 437; Robust standard errors in parentheses
* $p < 0.05$, ** $p < 0.01$

For example, in Table 3.2, I present the estimated probability of a vote against the Clean Air Act amendments given different levels of employment in the organic chemical production industry in congressional districts. In a state where at least one senator voted against the legislation, and the number of jobs tied to the organic chemical industries was at most 100, the odds of a negative vote by a member of that state's congressional delegation was about 10 percent. If the support by the Senate colleague was taken away, the odds of a congressional "no" vote when employment is no more than 100 is 2 percent. But in a district with 1,000 jobs in this industry and the support of at least one of the state's senators, the odds of a "no" vote at the congressional level increases to 17 percent. And under these conditions, with a minimum of 5,000 employees in the industry, the estimated chances of a "no" vote by the local congressperson was 62 percent. Even when the senators of that state did not vote against the Clean Air amendments, the congressperson with 5,000 constituents working in the CFC industrial sector had a 33 percent chance of voting against the ozone-protecting legislation.

There are, of course, a multitude of possible reasons that could compel a congressperson to vote against efforts to preserve our ozone layer, and my model does not account for these alternative explanations. At what might seem a very basic level, however, this evidence makes it clear that elected officials tended to act on what might be

Table 3.2 Chance of Voting Against the Clean Air Act Amendments by Size of Organic Chemical Industry Employment in District (percent)

Size of Workforce	*Probability of Voting "No" with Senate Support*	*Probability of Voting "No" without Senate Support*
Zero employees	10	2.0
100 employees	11	2.5
1,000 employees	17	4.3
5,000 employees	62	33.0

Predicted probabilities derived from results in Table 3.1

thought of as private incentives, that is, to try to ensure jobs in their districts, even if those jobs came at the expense of the ecosystem. At the time of the vote, there might have been a sufficient level of uncertainty for some to wage constituent employment against an uncertain impact on health and habitat viability. And a congressperson might rationalize that the payoff from protecting the ozone layer was too distant relative to the immediate cost in terms of jobs. However that internal logic was to play out, the outcome is consistent with a private incentive to defect from the collective effort to protect the ecosystem, and this has a corollary in the CO_2 debates that we confront today.

Learning from the Ozone Debate

This book is about the politics behind the regulation of processes by which humans are causing climatic patterns that put us at risk. The extent of that risk has been clear by the United Nations report on the impacts, adaptation, and vulnerability to climate change (2014). For the most part, the world community was successful at addressing the ozone hole, and my point in this chapter was to use what we learned in one collective action problem to help understand what we have to do in the next, and bigger, problem of climate change. Ozone production was banned in the 1980s, phased out in the 1990s, and by the beginning of the 21st century, the ozone layer is beginning to recover. So we have a critical success story from which to learn and build (Sandler, 2004; Berhauer, 2013).

The similarities between the ozone problem and that of climate change are that both are a function of human consumption, both required international and domestic regulation, both held the prospect for catastrophic consequences if no action were to be taken, and both levied a cost on the pursuit of taking remedial actions. The sequence of events that led to global action with regard to the ozone-depleting CFCs involved science and the initial link between CFC gases and ozone destruction, refutation and resistance by companies with an interest in continuing the status quo policies, and ultimately, the recognition that we had created a significant and measurable hole in the ozone layer that protects us from the UV rays of the sun. We were clearly harming ourselves, and at that point, resistance gave way to the articulation of a clear political path toward implementable solutions. But this pathway suggests that we have to go to the brink before the future is no longer discounted relative to the stability we seek today.

One of the critical differences lies in how we produce and consume CO_2 relative to CFC gases, and this difference presents one of the significant political hurdles to overcome. As individuals, we produce CO_2 as a by-product via various forms of consumption. This difference in production and consumption makes the collective action problem a much bigger one. The fact that 50 percent of CFCs were produced by one company in localized environments is the main difference between CFCs and CO_2. CO_2 is produced universally by most of the planet's population.

At the level of the individual, a little bit of denial can go a long way in providing the rationale for each person to avoid contributing to the collective pursuit of a stable climatic foundation. Given the difficulty of overcoming the collective action problem in the best of times, doubt about links between the large cars or longer showers and climate patterns makes it easier to continue patterns of consuming the products that cause CO_2 emissions. And if the good that you get out of being an "environmentalist" has a payoff that is accruable after you have died, then why give up the big car and the climate-controlled house if you will never see the payoff?

Olson's models suggest that we never will. Free riding has a higher expected value, and every consumer can shirk responsibility without detection. Solving the CO_2 problem, then, will require legislated restrictions on how or how much can be consumed, similar to the way CFCs had to be formally restricted. Prohibitive legislation (Yaffee, 1982) requires Congress to act and also requires many elected officials to act against their own immediate self-interests. The entire edifice of the industrial world is tied to energy from the burning of carbon products, and with such widespread investment, few individuals or industries have an interest in making sacrifices. But the role of science and the recognition of an ozone hole accelerated a response, generated a constituency, and moved policy makers to compel action. Droughts, storms, floods, and other perceptible changes in the local weather could play the role of the ozone hole. The window under which change is necessary shrinks rapidly, and long-term recovery can be expected only over generations (Hamlet, 2011). Other ecological collective action problems, such as the depletion of wolf and whale populations, took just a single generation to recover from.

4

♦ ♦ ♦

WHALES AND WOLVES

Examples of communities, countries, or the world overcoming the collective action problem to provide for the common good are numerous. It is harder, though, to find analogous examples with the complexity of CO_2 emissions and climate change. International treaties often generate cooperative outcomes that are for the good of all, even when not everyone can see the immediate benefits from cooperating. For example, the Universal Postage Union (UPU) was created by the Treaty of Bern in 1874 and made the free flow of mail around the world possible.

Prior to the postal treaty any country wishing to have a postal agreement with another had to make a bilateral arrangement. A letter to Russia from the United States would require a stamp purchased in U.S. currency. If the letter originated on the East Coast of the United States, delivery would have been carried out primarily by an international steamer company and then the Russian postal service, not the U.S. postal service. The 1800s brought a wave of immigrants to the United States, people with families in their home countries. Phones did not exist; air travel was a century away; the letter was the only means of communication. A U.S. stamp, though, had no currency in Ireland or Russia. The Universal Postage Union put procedures in place to make communication easier. The UPU had three basic principles: 1) a flat rate for all international postage, 2) equal treatment of domestic and international mail by national postal authorities, and 3) each country retaining the money collected for its postage. If the flow of outgoing mail in one direction was greater than the return flow, the recipient country would have to pay more than the dispatching country. The foundation of the UPU, however, was that if a lot of immigrants were sending letters home, the responses would equalize the flow. If the problem seemed tricky, the solution was simple.

The Convention on International Civil Aviation (CICA) and the Antarctic treaty are other such examples. The CICA mandates a common form of communication between pilot and air traffic control, and among other things, it prohibits countries from charging a fee to fly through their airspace. These rules seem obvious today, but they were not always in place and not having them would have caused untold problems, as well as dangers, for air travel.

The Antarctic treaty, which banned nuclear weapons on that continent, was also simple on the face of it. Part of the nuclear standoff was made stable by the ability of a targeted state to retaliate if their enemy launched nuclear weapons. A missile could reach the Soviet Union from the United States in 30 minutes—a terrifying 30 minutes, but enough time for the targeted country to ready and launch a retaliatory strike. Placing nuclear missiles on the Antarctic continent would mean the time from launch to detonation could be reduced by half. This would create a strong incentive to "launch on warning"—which minimized the chance that a leader could verify missiles *had* actually been launched and that it was not a mistaken interpretation of some other event. If the command and control systems for nuclear war were error free, this might matter considerably less, but there were numerous reports of false launches and other nuclear incidents, so relying on an error-prone system was far too risky (Sagan, 2003). To increase the time available to make rational decisions, nuclear-capable countries agreed to a ban on placing nuclear weapons on the earth's pole.

Each of these treaties is an example where international cooperation trumped the private interest of each state and overcame a collective action problem, and did so for a number of reasons. First, the costs of compliance were not very high relative to the benefits.

The calculation of an expected value, $E(X) = ((V_i * q_i * p_i) - C_i)$, was easy to see as being positive. Second, the largest group, the individuals in these countries, if they were even aware of nuclear targeting and deterrence, would have valued stability over instability. And third, in each case, certainty about cause and effect would have been high, and the future was not discounted relative to current benefits. But these conditions do not hold for all collective good issues, and certainly not for the thorniest of them, like climate change.

Saving the Whales

The International Whaling Commission (IWC) exists to regulate the commercial hunting of whales by individuals, firms, and nations. Two of the United States' staunchest allies, Japan and Norway, do not abide by the terms of the ban on commercial whaling and are, in fact, attempting to overturn the international agreement (Carlarne, 2005). International whaling agreements have been discussed and negotiated since the early part of the 20th century, but the first agreement with some level of coherent compliance by the member states of the IWC was the 1986 moratorium on killing whales. The formal ban on hunting whales is well within the lifetime of some people reading this book.

Whales are intelligent and peaceful; they care for their young for as long as humans do, and they are an integral part of the oceans' ecosystems. A whale washed up and struggling for breath on a beach makes the national news broadcast, because people care about their welfare, and whales are a public interest story. Whales were not always thought of this way.

In some parts of the world, whale meat is a dietary staple, and eating whale is a cultural custom that carries meaning in their collective rituals. The native peoples of North America and Russia are the most notable of these groups, and the IWC takes their cultural needs into account. Japan and Norway, too, claim that cultural imperatives demand that they permit whaling. Even though Norway has, for the most part, adhered to the moratorium of the IWC, Japan continues to grant permits to hunt whales under the guise of scientific research. It is estimated that Japanese whaling companies have killed 8,500 whales since the start of the IWC moratorium, and, as of 2005, 540 minke whales annually. This so-called scientific research supports the commercial market for whale meat in Japan (Carlarne, 2005; Friedheim, 2001).

Melville's great novel *Moby Dick* weaves a tale of the hunted and the hunter on the open ocean. In that era, the odds might have been slightly weighted against the whale, and the rate of culling whales might even have been sustainable over the long run. Whaling was once such big business that remote whaling stations dotted the oceans. The shipwrecked explorer Amundson found refuge in a whaling station on the tip of an island that most of us couldn't find on a map, and it was this remote whaling station that saved him and his crew. Whaling was part of our history, because whale oil was used for lighting lamps and lubricating machinery, bones and baleen were marketed for a range of products from combs to hoops for dresses, and whale meat was a source of protein. Living without the whale hunt would have been difficult at the time. But coal, oil, and electricity ushered in a new era, and whale oil lost much of its importance. This coincided with a decline in whale populations, making hunting more challenging. In 1833, the catch for Arctic whales was over 1,600. By 1900, fewer than 20 whales were taken (Ross, 1984, p. 13). Arctic whaling grounds had been overhunted and could no longer sustain a commercial industry. The American whaling fleet of 1846 was 736 vessels strong; by 1901, it was down to only 40 ships. And at one point, the whaling industry in America employed up to 50,000 sailors and although fortunes were made, Americans were loath to make whale meat part of their diet (Dolin, 2007; Ross, 1984). For those in North America, whaling was little more than an economic endeavor devoid of cultural ramifications, so when the economics were no longer efficient, and substitutes for whale products were already available, the practice was easy to give up.

In the days of the wind-powered whaling schooner, it might take a day to find, harpoon, exhaust, and haul in one whale, and the risk to those hunters was consequential (Dolin, 2007; Ross, 1984). The American and British whaling fleets managed to cull the North Atlantic whale population to the point of putting its sustainability at risk, and by the early 1900s, the industry on this side of the planet was but a shell of what it once was. That is not the story of the south Pacific and Antarctic whaling industry. The factory ship was introduced in 1925, and by the late 1930s, 50,000 whales were killed annually, mostly in the south Pacific. With the North American whaling ground depleted and an unsustainable kill rate in the Pacific, the international community began the process of negotiating an international agreement to at least manage the whale stocks. Extinction of the whale had become a possibility.

The initial intention was to set up a system to regulate whaling so as to make it commercially viable and sustainable over the long term. The International Whaling Commission (IWC) was formed in 1946 at the end of World War II and at a time when the United States was the dominant international player. The IWC's management programs were unable, however, to slow down the rate at which the oceans were being depleted of whales. The peak years of whaling—in terms of the size of the harvest—were from the 1930s through the mid-1960s. But by 1982, with the whale population at serious risk of collapse, the IWC passed a moratorium on whaling, and this moratorium was to be fully enforced by 1986.

The world community came together to ban the hunting of whales, because it was clear that the long-term cost of depleting the whale population was their extinction. Extinction of the whale population put at risk the entire ocean ecosystem on which so much of the human population relied. This was ethically as well as commercially undesirable, and as an example of collective action for a public good, this could be one model for thinking about the regulation of CO_2 emissions.

The moratorium on whaling did not, in fact, result from like-minded leaders of countries working toward saving whales from extinction. We can learn a lot from the global attempt to save whales, or at least certain whale species, but not necessarily about how global harmony works to achieve this end. In fact, Olson's (1965) argument provides a picture that describes why and to what extent the IWC moratorium works.

Olson's logic of collective action points to, among other factors, the size of the group as one determinant in the efficient pursuit of a collective good. As of 2012, there were 89 countries with membership in the IWC, making up roughly half of the independent countries of the world. By most standards in the international arena, this is a small group, and by Olson's logic, this group should be able to achieve at least some level of the provision of the collective good. And it has. The moratorium on whaling has reduced the annual whale harvest from 30,000–50,000 whales to a number much closer to 2,000–3,000 (Freidheim, 2001; Frances, 2012), a reduction by an order of magnitude in the annual kill, but not to the optimal level of zero whales killed in a year. The question is how this seemingly cooperative body generated an outcome that *may* lead to a sustainable level of whale harvesting in spite of having only a 50 percent participation rate among countries in the international system. If the impact of certainty and the future discount are considered at unity (p and q = 1), which seems reasonable when considering the state of the whale population in the 1970s and 1980s, the outcome must be driven by the expected value as determined by costs and benefits from saving whales from extinction.

Whaling in the mid-20th century was unsustainable, and evidence shows that antiwhaling efforts have been successful in recovery of whale populations. Keeping up the pace of killing evident in the 1930s–1960s would have led to complete collapse of the whaling industry, as well as the extinction of some species. From that perspective, all whaling countries had an interest in some level of regulation of the resource. Elinor Ostrom (1999) would put this in the category of a rivaled collective good, a common resource for which a small group should be able to self-regulate its use. In this area, there is not a lot of tension between Ostrom and Olson. The rules of the IWC and its whaling moratorium illustrate how cooperative behavior might inform approaches to achieving success in the climate change arena.

The IWC whaling moratorium is an example of a successful effort to overcome the collective problem of overexploitation of the oceans. People and their political leadership recognized the problem, and after numerous attempts to control the destruction of the whale population, the world community implemented a complete ban on whaling. The ban still holds to this day. From Olson, we would learn that it takes small groups who stand to benefit in proportion to their commitment to securing the good, and if the group is large, selective payouts might be necessary to generate participation. Ostrom's outlook is that a group with a collective interest in managing public resources should, under some conditions, be able to self-regulate their individual incentives to exploit the commons.

The rules that guide the IWC show a different view of how the world community comes together to protect whales but may reveal ways to save the climate in spite of individual incentives to ignore CO_2 emissions. Membership in the IWC is voluntary, and the institutional rules that govern decision-making require three-fourths approval of any regulation. So a super majority is necessary to bring into force a ban like that on whaling. But because membership is voluntary, no enforcement mechanism is built into the IWC charter. If a country chooses to violate the regulations, enforcement is only effective relative to what individual states would impose on the violator. About half of the member states of the IWC are small island nations that carry no weight when it comes to individual enforcement efforts, but the other half are larger and more industrialized countries that could impose costs for noncompliance. This means that the majority could impose rules on the behavior of sovereign minority states, if they have the wherewithal to make that imposition work. This is what Olson meant when he referred to the instrumental role of an oligarchical relationship in increasing the chances for securing a public good.

In our sovereign state system, it is difficult for nonenforceable agreements to compel behavior from member states, and the IWC recognizes this. The three-fourths voting rule creates a majority system that can violate the preferences of the minority, who in this instance are sovereign members of the international system. IWC has two mechanisms for the minority to respond. One is to file an objection; the other is to withdraw from the IWC (Carlarne, 2005). Filing an objection is the way a state puts the IWC on notice that it does not feel obligated to abide by the terms of the moratorium, as both Japan and Norway have done. They reserve the right to engage in commercial whaling, and Japan acts on this with an aggressive "research program" that takes over 1,000 whales per year.

Certainly, there is public angst at Japan's continued hunting of whales. The TV show *Whale Wars* shows the struggle between those opposed to whaling and the Japanese claim to the right to hunt whales. The majority is unable to enforce the rules on Japan. Sanctions don't work well, because the target is a strong ally, has the world's third-largest economy, and is a major trading partner. Over the years since the IWC moratorium, the United States has threatened to sanction Japan and Norway but never carried out the threat, and in recent times, the threat is ignored (Carlarne, 2005; Friedheim, 2001). In addition, many of those majority countries make exceptions to the moratorium to take into account their native populations. Canada, Iceland, Greenland, the United States, Norway, Japan, Russia, and others all claim to have native peoples for whom whaling is a cultural custom, and the IWC makes provisions, even quotas, for these groups. These provisions, however, are not made for the Japanese as a group (Friedheim, 2001). There is, of course, considerable ethical and political debate on this issue.

There are some remarkable things about the whaling moratorium, not the least of which is that it has given many whale species time to recover after they had come close to extinction because of mechanized hunting and processing. In that sense, it has been successful. What the whaling moratorium shows us is that 1) cooperation is possible even in the face of dissention, 2) powerful members have a larger call on shaping the outcome than weaker ones, 3) a public campaign presenting evidence can be effective in turning the tide of the political debate, and 4) when costs are stark, everyone can see the downside outcome, and even if they disagree with the prescriptions, they will be more selective in violating the norms. These observations have implications on our collective ability to confront the continued anthropogenic contributions to climate change.

From Saving Whales to Saving Our Climate

Whales in the world's oceans and a stable climate are both collective goods, and in both cases, securing cooperation in the face of individually rational interests has posed problems. Today, however, enough whales are swimming the oceans that a tourist industry thrives around them, and this is a significant and positive reversal of the whale's fortunes over a reasonably short period of time. Such a reversal is imperative for the problem of carbon emissions, and the case of whales has something to teach us about how to go about achieving it.

Sometimes, dramatic events or evidence can change the way we think about costs and benefits. If the new information is sufficiently wrenching, it can make it difficult to deny causal processes and discount future benefits. When the ozone hole appeared in the Southern Hemisphere, it became hard to deny that something drastic was in motion, just as the scientific recognition of the link between smoking and lung cancer changed the way we see that private action of smoking. The decimation of the whale population made the cost of continued commercial whaling painfully stark, and even those who saw whaling as a sovereign right and for whom whale meat provided a significant amount of protein for their population could see the costs of doing nothing to prevent the populations collapse. Japan and Norway did not want a complete ban on whaling, but neither country has returned to the massive commercial whaling of its past. Japan may be violating the IWC moratorium if their scientific research can be seen as such, but this violation is relatively modest compared to the 27,000 whales they killed in 1965 (Friedheim, 2001). Japan could violate the moratorium by whaling at the level of their peak harvest, but they choose not to. This could be seen as progress in the fight to save the whales.

Whale hunting has a parallel to CO_2 emissions in the sense that both are a function of consumption rates. The consumption of large quantities of whale meat dictates the killing of a large number of whales. The consumption of large amounts of fossil fuels for lighting, transportation, and industry will result in the production of large quantities of CO_2 that will adversely affect our ecosystem. At the end of World War II and through the 1960s, whale meat was the most prevalent form of meat-based protein in the Japanese diet. Japan, and other countries, in the face of the complete collapse of whale populations, restricted their consumption of whale-based protein to a "necessary level." The alternative, with the disappearance of whales, would have been complete removal of whale meat from their diets. You could make an argument that the longer Japanese society lives on reduced levels of consumption, the more likely they are to continue to reduce whale protein as a cultural and dietary supplement. Cultural adaptation may be one result of a global ecological challenge (Wilson, 2002). A movement away from a reliance of fossil fuel consumption, if sustained, could result in a similarly declining trend in the ways and amounts of our CO_2 production.

A second lesson to learn is that when the issues are political, the public matters. To a significant degree, the public ultimately pays the cost of change, and therefore once there is enough public pressure, the political dynamics provide greater latitude to discuss remedial strategies. In the whaling debate, much of the world saw whales as a creature that was much closer to us than to a fish. Their brains are large; they communicate, and they nurture their young as humans do with theirs. *Free Willy* popularized the idea that whales are smart and loveable and imposed a social cost on those who thought of marine mammals as merely a protein source. In the United States—and frankly most of the Western industrialized countries—the public sees whales in this more-intelligent-than-fish perspective, and it is no accident that the United States and these other Western countries dominate the IWC meetings and prevent the breakdown of the whaling moratorium. Then, too, whale meat is not a component of the dietary intake of most of these countries. Once lamps could be lit with petroleum or electricity, and once whale bone was replaced by synthetic products, killing whales had little point. This perspective is very different from the Japanese one, and public opinion from both sides influences the political debate and the discussion about remedial strategies. Those in the nonwhaling countries, though, have more influence.

In his best-selling book, *Cosmos* (1980), Carl Sagan asked how we would respond if we discovered a planet with a life form like our whales. According to Sagan, we would be ecstatic. We will have explored millions of miles in search of life in the universe and found a living being that talks, remembers, and sings, indicating a long evolutionary process for the development of intelligent life. Sagan pointed out that what we are in search of in the universe is what we are killing here on our own planet. Sagan's observation predated the IWC moratorium, and Sagan also produced one of the most highly viewed TV documentaries of the time that portrayed our universe and our planet as thriving ecosystems. Public attitudes were, in part, shaped by the idea that what we search for in the heavens might be what we were killing in large numbers on Earth. The costs of the continued killing of whales became too much for the public to bear, and the potential payoff from a ban sufficiently high to move public attitudes. In Japan, this story would be significantly different.

The story of our collectively saving the whales also demonstrates that power matters in the provision of a collective good, just as Olson would have suggested. Japan, Norway, and a few other countries still disagree with the moratorium on whaling, and Japan actively resists full compliance. But Japanese whalers do kill considerably fewer whales in a given year than they have in decades past, and this appears to be a result of the international moratorium, not their national understanding of the role of whales in our global ecosystem or in our lore. Japan succumbs to the collective preference of the more powerful coalition of countries, but this might only be a function of the relative influence of those antiwhaling countries. As Japan has become more economically important in international commerce, so too has its influence over smaller countries, such that today it tries to motivate or entice members to join the International Whaling Commission to get the votes necessary to overturn the moratorium. The power to influence international agreements seems to matter a lot.

The moratorium on whaling and the rules of the IWC also point to the conclusion that unanimity is not necessary to achieve some level of provision of the collective good. Although a complete and enforceable ban on whale hunting does not exist, the whale population has stabilized sufficiently to allow the idea that regulated harvesting could be sustained and should be permitted (Friedheim, 2001). The imposition of a moratorium by the most powerful countries and the compliance, or at least partial compliance, by most of the rest of the world demonstrate that part of the ecosystem can be stabilized as long as consumption, including carbon consumption, can be reduced to a level that is sustainable over the long term. Humans *qua* humans did not in themselves create the climate problem we face, nor was CO_2 production always a significant problem during the human epoch on this planet. CO_2 consumption became a problem when the rate of consumption of carbon-based energy outstripped the earth's ability to compensate. This is a consumption issue related to, among other things, population growth. But there is nothing inherent in either the human species or the available technology that requires us to produce CO_2 in quantities that are unsustainable. This is something that we choose to do, if by no mechanism other than our inability to confront the collective action problem through our political processes. The ozone issue was dealt with when certainty, time, and expected benefits came together to make it an easy choice. The largest creatures on our planet were brought back from the point of extinction through cooperative and collective behavior. Likewise, wolves still roam the lower 48 states, even if through a different mechanism.

Saving Wolves from Extinction

I grew up in the state of Michigan in the 1960s and 1970s, and I spent much of my young adult life camping, hiking, and canoeing in the forests of the north. I always hoped that there were wolves out there in the deep woods, but I never saw one. I know now that there were, in fact, no wolves in Michigan from the early 1970s until the late 1980s. Bounties for wolf kills were still in place until the mid-1960s, and human population movements restricted wolves' range. The wolves of Wisconsin and Minnesota were close to extinction as well. Wolves were eliminated from the North and Southwest, by 1925 in the Arizona and New Mexico region and by the mid-1940s in Montana, Idaho, and Wyoming (Brown, 1983; Theil, 2001; Wydeven et al., 2009a).

In spite of this, wolves are back, and most people seem to think that there is value in having them as part of the natural system in America. In fact, studies show that the American public is willing to pay $67 per person annually to maintain the wolf population, so there is monetary value as well in securing the habitat for wolves (Loomis and White, 1996).

In 2012, the National Park Service reported that there are roughly 1,800 wolves in the northwest region from Wyoming to Washington and Oregon. The wolf population in Minnesota was estimated to be nearly 3,000 individuals in 2008, the Michigan population at greater than 400 (Minnesota DNR, 2012; USFWS, 2012; McWhirter, 2011). These wolf populations have risen from extinction or near extinction to thriving and stable. If we consider that there is a collective good in rescuing wolves from extinction in the continental United States, Olson would tell us that there has to be a net positive gain for the individual and the group. I would argue that there had to be reasonable certainty about the consequences of not doing so. There would also need to be a low discount for the future relative to today. To go back to the equation that shapes this argument, $E(X) = ((V_i * q_i * p_i) - C_i)$, q_i and p_i have to be reasonably high, C_i pretty low, and V_i high. People derive great pleasure from seeing bears, deer, snakes, and of course wolves in the wild. But most people are not at risk of losing their animals or pets to wolves, or other wild animals either, so the cost to them for providing a secure habitat for the wolves is minor. This is not necessarily true for the rancher or the sheep farmer. This debate has nuances that, say, the ozone issue did not have, making achieving success more difficult. And yet, success has been achieved. Wolves are no longer in danger of becoming solely a mythical creature.

If wolf recovery is not quite the collective good that ozone and climatic stability is, it is useful to think about how we protected an endangered species without that overwhelming sense of urgency being felt by most of the population. We might be on a quest right now to save another species from extinction—humans—and yet there is no pervasive sense of urgency (p or q). How, then, were wolves saved when many might not have known that the wolves were endangered, or even that their rescue was an important factor in their lives? Along with spotted owls, bald eagles, grizzly bears, snail darters, and a collection of minnows and insects that most were thoroughly unfamiliar with, wolves were saved from extinction by the Endangered Species Act (ESA) of 1973. The United States signed and ratified international treaties that put this process in place, particularly the Convention on International Trade in Endangered Species of Wild Flora and Fauna (Yaffee, 1982). The ESA followed a trend within Congress to define and defend endangered or threatened species, to which the ESA added breadth and enforcement mechanisms (Peterson, 1999). But at its core, the ESA was a domestic creation with widespread support among the lawmakers of this country in that era.

The ESA is thought of as one of the most intrusive laws ever passed by the U.S. Congress, in part because it subjugates states' rights to those of the federal government, and importantly, it leaves little recourse to challenge based on economic efficiency or the relative impact of saving one small animal over the needs of the broader population. For example, the Tennessee Valley Authority (TVA) had to confront the dictates of the ESA and halt the construction of the Tellico Dam because completion would have threatened the snail darter fish. The Supreme Court upheld the rights of the environmentalist—and the snail darter—over the TVA, demonstrating the strength of the act in preserving biological diversity. As a result of this court decision, Congress amended the ESA, and the Tellico Dam was finished, but the markers were laid down, and endangered species would have their day in court (Peterson, 1999; Yaffee, 1982). Just to put this into context, the snail darter is part of the perch family and is only 2–3 inches in length. A small fish turned a rather large dam project into a national scandal, and at least in the short term, the fish won by virtue of the Endangered Species Act.

In spite of the draconian implications for economic expansion and the life of plants and animals in the United States, the ESA was not a highly contested piece of legislation. The House version of the bill passed with a vote of 390 to 12; the Senate version received unanimous support, and the version that came out of the conference committee had four dissenting votes in the House. No one spoke on the floor against the legislation. The version of the legislation that was passed by both houses of Congress was initiated by the White House under President Nixon. We had a Democratically controlled Congress and a Republican White House, and both felt strongly enough about protecting endangered species to articulate, pass, and sign intrusive legislation almost without dissent. Yaffee calls this a "prohibitive policy" in that it prohibits actions that put at risk other outcomes without balancing social or economic factors (1982).

It is easy to find reasons to think of any species on the planet as a public good, and most as a rivaled good. A nonrivaled good is one in which consumption by one group or individual does not affect the consumption of any other; air and water are generally in this category. A rivaled good is a bit different, because for each unit I consume, there is one less of that species (or other good) for you to consume. Fish in a lake are a great example here. Because nobody has an individual incentive to restrict his or her fishing, overfishing is the result. Overfishing, overhunting, or overconsumption of a rivaled good makes extinction a real possibility. In the 1960s and 1970s, environmentalists began to shape the debate in the United States over the right to push a species to the point of extinction, and at that time, the overwhelming majority of Americans, and certainly almost every representative and senator, thought action to protect the endangered had to be taken.

In a number of dimensions, there is real risk to humans in allowing plants and animals to go extinct. Medicines, for example, often come from plants. The cure for diseases incurable today may exist in plants that are allowed to die out. We may be discounting the future value of a plant without even knowing that value. With top predatory animals, the consequences of near extinction are often immediate. DDT affected reproduction in peregrine falcons, and the population in New York State plummeted to near zero. With the ban on DDT, by 2009, there were at least 17 breeding pairs in New York City alone; people could watch them from the Empire State Building, and falcons feed on pigeons and other feathery prey so the city bird population remains in balance (Subramanian, 2009).

Photos of river basins in Yellowstone Park before wolves were eliminated from the western states show aspen trees on riverbanks thick with foliage. After the top predator in the ecosystem was gone, foliage was sparse and unhealthy. The deer and caribou, once hunted by wolves, now had time to pull all the foliage off the trees before they moved on. The foliage returned with the wolves, and photographs of before and after clearly show this. There is evidence that the reintroduction of wolves in the 1990s led to the reinvigoration of the ecosystem, because caribou that got lazy when there was no predator to run from had to be on alert when in open river basins (Kauffman et al., 2010). This natural inclination to be alert was removed with the wolves, and their return facilitates a healthier caribou population that can thrive (Laundre et al., 2001).

With environmentalists in charge of shaping the debate and the clear impact of not helping those species on the endangered species list, the media became an instrumental tool in the political process of shaping our national concern for some plants and animals that we didn't know existed (Yaffee, 1982; Peterson, 1999). Those making a case against protecting animals and plants had an awkward argument. The only significant opposition to the ESA, and this opposition was relatively muted, was the fur industry. But the argument to hunt a species to extinction for fur was not a good policy position even in the short term, because doing so would put furriers out of business. The constellation of forces pushing for the adoption of the ESA was so overwhelmingly one sided that the politics were "easy."

Saving wolves and snail darters does not, of course, have the scale or magnitude of CO_2 emissions and climate change. Few know of the snail darter; everyone drives a car. Saving the spotted owl is not a priority for most people; cooking meals and heating homes are. Wolves and whales are an important part of our national psyche and ecosystem, but a carbon-based industrial sector is at the foundation of the system upon which advanced economies depend. Still, wolves, whales, snail darters, and spotted owls are examples of victories in overcoming the collective action problem, and elements of these victories can be used to influence the climate debate. A number of points stand out.

First, the cost associated with remedial actions for wolves and whales was not high, at least to the average citizen. There was also significant value to protecting wolves, which was made clear by the well-organized environmentalist movement. The public campaign played a critical role in mobilizing support, and the image of the country's top predator in danger of turning into myth and legend, even for people who never camped in wilderness, amounted to tragedy.

From the public's view and that of Congress, the value came with little apparent cost, so the number of free riders was minimal. Wolves were a public good that almost everyone thought was worth keeping.

Second, uncertainty and the future discount rate were very low when it came to saving wolves. In fact, wolves were not endangered in all of North America, only in the 48 contiguous states, but the possibility of losing wolves was not disputed. Wolves had been eliminated from the West for years and the Great Lakes–area population was on the verge of becoming a historical legend. If wolves were not protected, they would be gone forever. Along with wolves, other endangered species were protected. At the individual level, you could discount that future, but there would be no wolves in it, and there was very little option but to accept the inevitable elimination from the lower 48 states. If we wanted the wolves to be spared, we had to act (Mech, 1995). And with the wolves came the snail darter, the spotted owl, and a host of other plants and animals that most were unfamiliar with.

The ozone hole helped shape and change public awareness and opinion about the atmosphere; whales did the same for life in the oceans, and the wolf for endangered species. Apart from being crises in themselves, they were all highly symbolic, illustrating bleak futures and prospects of bad things to come. In this sense, the wolf may have been a signaling mechanism that raised awareness of the linkages between our actions and the health of our ecosystem. It is this awareness that might help with the climate change struggle.

Finally, the role of environmental groups and the media in pushing the ESA legislative agenda (Peterson, 1999; Yaffee, 1982; Schanning, 2009) is undeniable. Given the magnitude of the changes in the relationship between development and our ecosystem by the ESA, it is remarkable how little resistance was encountered. A Republican president effectively initiated a piece of restrictive legislation that won nearly unanimous support from both houses of Congress, which were controlled by Democrats. There either had to be a very strong normative dimension to the problem at hand—permitting the extinction of species at a rate unparalleled in human history—or there was little political cost to going with the flow. This strong sense that something needed to be done and widespread protection of endangered species was the right thing to do must have been pervasive. One version of this story is that politicians focused on a normative imperative rather than political expediency, something that seems to have been inverted in the climate change debate. The private interests now trump the public good, as Olson would describe, but there was a point in our political dialogue where the small group reached a socially optimal outcome to a vexing social problem.

The decimation of wolves is, in many ways, more like the problem of depleting the ozone layer or killing whales than the issue of carbon consumption. Wolves were hunted by a few people, and state governments provided bounties for each wolf taken up into the 1960s. A conscious policy came up against the outcome of eliminating one of the world's great creatures. Wolves, from Jack London's novels to fairy tales, are a part of the American psyche that was about to become myth and legend because of state policy and the preferences of a few individuals. The ozone hole as well as the decimation of wolves and whales were caused by human beings being either permissive or industrious. Carbon dioxide is a molecule that we create by living, in the normal course of our daily lives, as we heat our homes, cook our meals, and drive our cars, so the costs are distributed more widely, and yet the consequences will be more drastic. The examples of wolves, whales, and even the ozone hole are relatively less consequential, but the central parts of each of these efforts will serve to craft a strategy that will change our behaviors.

5

♦ ♦ ♦

MINIMIZING UNCERTAINTY AND FUTURE DISCOUNTING

To make sense of how we make choices with regard to climate change, it is important to consider the implications of inaction—of doing nothing—to change the rate at which we emit CO_2 into our atmosphere. Thinking about the implications of doing nothing in the face of climatic pressures feeds right into the question of whether those of us in position to do something will even be alive by the time any remedial changes would be perceptible. Asking those questions, however, begs for something of a dire description of possible outcomes, some of which I feel compelled to provide. Writing a chapter on possible consequences from not acting in response to impending climate disruptions allows me to be something of a screenwriter for a movie house, able to portray the dramatic with the expectations that it is all fiction anyway. Unfortunately, it is not fiction, and at best, what I can describe is something between an empirical reality and a future forecast. As a political scientist, forecasting future climatic conditions is not in my realm. The work of climate scientists, though, makes it possible to paint a clearer picture. The actions for reducing CO_2 emissions are based on the premise that climate change will wreak havoc and require major adaptation strategies (Javeline,

2014). Responses vary, and doing nothing as a strategy at least requires rational adaptation. For example, the Dutch, who are famous for the levee system that prevents flooding, have begun to allow in the ocean water. Accepting the rising oceans as a new reality, that is, no longer trying to fight rising oceans, is their adaptation strategy. In the United States, we are building taller levees. Doing nothing as a strategy for mitigating our impact on the climate at least requires recognition of the future consequences so that our actions can be judged relative to alternative expectations.

The Cost of Inaction

If our willingness to discount future benefits is at the heart of collective inaction on climate change, one way to generate movement is to develop a clear understanding of what that future might look like. Portraying the possible outcomes of inaction in the starkest possible terms is one way to do this. There comes a point at which the costs are so immediate and the consequences are so near that, beyond that point, any meaningful change is moot. As long as we can discount the future benefits from taking action to mitigate our impact on the climate, the model for overcoming the collective action problem doesn't have a payoff that makes participation worth the effort. If you view this from the perspective that any benefits are three or four generations out, you might conclude that there seems to be little point in paying a cost today, because humans tend not to know their offspring in the fourth generation. When climate consequences wash away homes and destroy neighborhoods though, the expected payoff from taking action, or the fallout from taking none, is more immediately felt.

If global warming and climate change are just part of the natural cycle and there is nothing we can do about it, then the "do nothing" strategy is perfectly rational, and there is no choice but to accept new coast lines, superstorms, and everything else nature throws at us. My argument is, of course, predicated on the conclusion that changing weather patterns and the increasing temperature of the planet are, in fact, directly tied to the overconsumption of carbon-based products.

Understanding the potential consequences of these behaviors is important because they shift the discount rate.

There are precedents for linking the realization of dire consequences and action to change the status quo. The hole in the ozone layer over Antarctica was one such event. Faced with clear evidence that consequences were upon us today and not projected for some future time quickly brought on remedial action. People in New Zealand and Australia understood that the consequences would happen not to far-removed generations but to them, during their own lifetimes. This realization changed their daily behaviors but also was part of moving the international community to ban ozone-destroying gases. The level of certainty about cause and effect was brought home in stark fashion, the value of the future benefits, p, suddenly became close to 1.0. Within months, those legislative changes required to repair the atmosphere were in place.

If the political struggles to do something about the increasing and deleterious changes to our climate are rooted in the ease with which critics of an anthropogenic explanation can spin an argument that prevents national legislation from moving us forward, maybe we have to be more imperiled than we generally think we are (Davenport, 2014a). If we had a hole or a smoking gun for climate change, maybe more of us would be willing to rethink our patterns of consumption. But for now, most of us prefer to have our DVR player always at the ready and consuming as much electricity as a refrigerator so that it will turn on at the touch of a button. A "climate hole" might compel us to rethink our need for convenience over our need for a stable planet.

One way to think of a doom-and-gloom scenario may simply be a series of events that are the equivalent of the ozone hole, the decimation of the whales, or the realization that animals that go extinct aren't coming back, and when we let these things happen, there are consequences. So a depiction of the harrowing, of the potential changes wrought by our overconsumption, might be our ticket back from the brink. And it is with this in mind that I describe some of the consequences that we potentially face. As Coral Davenport (2014a) of the *New York Times* put it, if we cannot get this under control, the "planet will be locked into a dangerous future."

As a political scientist, I am reluctant to predict future outcomes, but climate science and global circulation models help detail possible scenarios. These bring home, in a very real way, what climate science has predicted will happen if strategies to mitigate global warming are not put in place. The hope is that, like in the ozone debacle, knowing the consequences will reduce uncertainty, make it harder to discount future outcomes, and drown out the critics who so easily spin arguments that prevent national legislation from moving forward. Knowing the direst consequences may allow us to experience the immediacy that was instrumental in allowing rapid political action in the ozone case.

A Set of Possible Alternative Worlds

In spite of the failures at Copenhagen, two outcomes from the summit allow us to judge the likelihood of facing irreversible and dramatic weather patterns. The first is President Obama's commitment to a 17 percent reduction in GHGs by the year 2020. This is 1 percent short of the goal the House of Representatives approved in HR 2454, but it does provide a pathway to reductions in line with movement toward the Copenhagen target of 50 percent by 2050. The second outcome of the summit was a vague international commitment to keep the rise in global temperature within a 2°C limit. If the world community thought that a rise in global temperatures of only 2°C was sufficient to stave off dramatic changes to our climate, then a look at what that world holds should give us some sense of things to come. Individuals will at least be able to judge their quality of life if we can hold the world to this 2°C commitment. From that point, we can also then ask the question of what the world looks like if we miss that target temperature, which would be consistent with current trends. We should also keep in mind that this commitment of a 2°C increase limit is an indicator of expectations for global temperatures to rise considerably more than 2°C—that is, a 2°C rise is a better outcome than what is expected if we do nothing. And to put this into perspective, the U.S. Environmental Protection Agency estimates that global temperatures could rise between 2°C and 12°C over the lifetime of a baby born today (EPA, 2013a). The variance is huge and a result of uncertainty in climate modeling, but a temperature increase anywhere between 2°C and 12°C will result in dramatic changes to our physical and social world. The low end of this estimate is likely to occur even if we take aggressive steps to change our CO_2 emissions; the high end becomes more likely if nothing is done.

To understand what our world would look like with a climate that is on average 2°C warmer than it is today, we have to rely on projections developed by scientists who study both past and contemporary climate patterns. Ice and land coring are some of the techniques used to understand past temperature variations of the planet. There is always uncertainty in the realm of science, but the uncertainty in these temperature estimates is measurable, and using estimates of past climate conditions provides a range of possible future conditions. Moreover, significant changes are observable on the planet today that point toward likely long-term outcomes. If the historical patterns line up with what we are observing in our contemporary world, we should take these models and their forecasts seriously. There is room to argue that these are only projections, or hypothetical situations, but the weight of evidence and argument aligns me with climate scientists and the projections and hypotheses they make.

To understand what a warmer planet holds in store for us, I will rely on a number of scientific projections, contemporary events, and other descriptions of possible outcomes. Mark Lynas (2008), for example, has presented a graphic view of what those two degrees will cost us in terms of our ecosystem, climate stability, and ability to adapt and cope. Two degrees, though it seems like a small rise, is in actuality a significant increase in global temperatures. First, the increase in the average temperature is on the Celsius scale. This translates into roughly a 3.6° increase on the Fahrenheit scale, which most of us are more familiar with. For example, the average high temperature in Miami in August is about 90°F, and the average low is just under 80°F. A 2°C increase would push Florida into the mid-90s during the day, on average, and the mid-80s at night. Some may see this as an asset in Florida. But that change in temperature in a different climate setting, in Anchorage, Alaska, for example, will accelerate the onset of spring and delay the onset of winter by weeks. It might prevent the freezing of the sea, which, in turn, prevents polar bears from finding food. With the decline of sea ice, polar bears could be brought to the point of extinction. A 2°C increase is a large change in global temperature, and this is an average temperature, with some latitudes and climates being hit significantly harder than others. The polar regions heat up faster than the midlatitudes, so if the planet increases by 2°C on average, the implication is that Alaska, the Arctic, the Antarctic, and Greenland will all be considerably warmer than the 2°C average. These regions also hold most of the surface ice that would melt and influence sea levels around the world. An average increase of 2°C is not as minor an increase in temperature as it seems. It would be best to think of 2°C as a significant increase, though one in which we could adapt and survive. This, at least, seemed to be the hope of the world leaders at the Copenhagen summit.

A 2°C increase in global temperature is one of many possible alternatives. The temperature could rise 5°C, or it could stay stable at present levels. These are alternative outcomes from which to evaluate the prospects for confronting what may come. Focusing on what might happen given the continued production of CO_2, we can judge the importance of uncertainty and the future payoffs from the different outcomes. The model that guides this book is based on an expected value $E(X) = ((V_i * q_i * p_i)-C_i)$, which requires thinking about the value and the costs in relation to the likelihood of achieving a desired outcome. These alternative worlds are indicators of the value and cost of acting now. We can also think in two metrics, as climate scientists do: one in terms of the temperatures we reach in a given year, 2050 or 2100; the other in terms of a degree change in temperature. Each will hold some level of the normal uncertainty associated with scientific understanding and prediction, and each will confront the question of whether we observe outcomes that are a result of natural processes or anthropogenic ones.

When focusing on the time element of climate change scenarios, age and life expectancy should be put into perspective. A college student 20 years old today would be in his or her mid-50s in the year 2050, possibly with children in college and looking forward to a retirement of 30 years. A U.S. government climate report projects that the average global temperature in 2050 under conditions where we reduce global emissions of CO_2 will be somewhere between 2°C and 3°C warmer than today (Karl et al., 2009). To put that into the more familiar Fahrenheit scale, the average temperature would rise in the range between 3.5°F and 5.5°F. In the year 2100, when climate predictions are the scariest and yet the most uncertain, few if any readers of this book will be alive. But our children probably will be, as will our grandchildren. Some projections for global temperatures for the year 2100 have the average 11°C warmer, roughly 20°F, than at present. The average August temperature in Michigan at the time of writing is about 80°F. If these projections are correct and we do nothing, an August vacation to Lake Michigan that your grandson takes with his children might be topping 100° in the shade.

The temperature in any given decade will be a result of small increases that add up over time. Global temperatures are rising now, and the question is only how fast these creeping increases will take place. If temperatures in 2050 will be 4°C warmer, that does not mean 2040 will have the same temperatures we have today and things will suddenly jump up over the final decade. Over the last 50 years, the average temperature in the United States has increased by 1°F, but over the next 37 years, it is anticipated that it will increase somewhere on the order of 4°F, particularly if we do nothing to change that trajectory (Karl, 2009). The rate of change, therefore, is roughly double. Importantly, the trajectory we are on, or the trajectory that predicts a possible outcome, is a function of the actions we take—or do not take—to overcome the collective action problem. Put differently, when climate panels describe possible pathways and rising oceans or wildfires associated with a particular level of global temperatures, they are implicitly pointing to the effects of the politics that underlie the climate process. We and our actions can alter the trajectory, and climate scientists have names for these likely scenarios—the A2 or the B1: the "business as usual" and the "go getter." The functional difference between the "business as usual" (A2) and the "go getter" (B1) scenario is the extent to which politics constrains or incentivizes our ability or willingness to consume carbon. One way to think about these alternative worlds is in terms of whether the costs for making the changes required to get on a better pathway are less than the costs of dealing with the consequences that result from not making any changes.

The Best of Possible Worlds

A 2°C average increase in global temperatures seems to be the best outcome we can hope for from the conditions we have generated. Even if we were to stop adding to carbon dioxide levels in the atmosphere, it would be decades before changes were evident, and the same applies to rising temperatures—the changes come on slowly. Most scientists and committees agree there will be at least a 2°C increase in temperatures before the end of this century. Because the polar regions will heat up faster, people in Alaska might react more forcefully to a 2°C global increase, while a person in Indiana might react more tepidly. A 2°C hotter planet won't always be fun, but for most, it will be survivable with moderate adaptations.

At 2°C, we will likely see an increase in dramatic weather events, some of which are occurring already. Superstorm Sandy lashing the East Coast and causing such widespread destruction was one such occurrence, and it was a new type of event for most people. Seawater flooding the subway system of New York City was never part of the debate; with Sandy, it became an unfortunate reality. Most projections for 2°C higher average temperatures show significantly more of these massive storms, and they are occurring in places not normally threatened by such events. Binghamton, New York, lies in the hills of the upstate region where the highest mountains are in the range of 1,800 feet. Binghamton is also at the confluence of two rivers, the Chenango and the Susquehanna, which carve out the valleys of the region. All rivers flood occasionally, and the larger rivers have proportionally larger-scale flooding. In 2006, a southern hurricane moved inland pouring copious quantities of rain on the hills of the Southern Tier, as they call the Binghamton area. The water drained into the valleys causing the two rivers to breach their banks. Hundreds of people flocked to emergency shelters because of the damage to and destruction of their homes. Some of these homes actually moved off their foundations, while others filled up with water, as much as four or five feet into the first floors. This was a small-scale version of New Orleans in 2005 but still catastrophic for the region.

The Federal Emergency Management Agency (FEMA) prepares flood maps for river basins around the country. These flood maps determine if flood insurance is required on homes, and if it is, how much it will cost. The most common metric for the FEMA maps is the 100-year flood line. The 100-year flood line is FEMA's estimate of a 1 percent chance of flood waters reaching that point in any given year, and on average, 100-year floods are rare events. In fact, if a home is outside the 100-year flood line, insurance companies do not see flooding as a significant risk and therefore do not require insurance against flood damage. The Binghamton-area flood of 2006 went well beyond the 100-year flood zones in many areas, so from this perspective, it was a very rare event.

In September 2011, two other significant storms spent themselves out over the Southern Tier during a 10-day period. The first one, Tropical Storm Irene, soaked the ground to the point that it could hold no more water. The next one, Tropical Storm Lee, discharged so much water in the hills overlooking these two river basins that the rivers continued to rise for several days. The flood that resulted from these strong tropical depressions went well past the 100-year flood zones, and in some neighborhoods, flooded areas considered to be in the 500-year flood zone. Many homes that survived the 2006 flood were completely wiped out in the 2011 floods. Of course, any low probability event is possible, just not very likely in any given time period. But two low-probability events in a short time period are very rare. Though these events may not be the smoking gun of climate change to anyone but residents of the Binghamton area, they are now part of the dialogue. A significant number of residents have left their damaged homes and moved away. They are what we might think of as climate change refugees, moving to locations where their family pictures are safe, their investment in a house is not continually at risk from rising flood waters, and their worries are about the normal things in life. Many dozens of flood-damaged homes stand abandoned in the Binghamton community. Many readers of this book may not know where Binghamton is, so the point to note about the location of Binghamton is that it is a few hundred miles inland from the coast and easily many hundreds of miles from what we might normally think of as a hurricane zone. It is also worth noting that this odd climate situation took place when the average global temperature was only about .5°C warmer than it was 50 years ago (Karl et al., 2009). A 2°C change in average temperatures is greater by a factor of four.

The disruption from a 2°C change upward in average temperatures would be widespread and certainly more consequential than a series of floods in the Binghamton river basin. Another notable example is the summer heat wave in Europe in 2003. June, July, and August tend to be the hottest months of the year in the Northern Hemisphere. But the summer of 2003 broke all records. France, Italy, and Spain saw over 20,000 die as a result of the heat wave, 11,000 of whom were in France alone (Beniston and Diaz, 2004). Some estimates put the number of deaths across all of central Europe as high as 70,000 (Robine et al., 2007). The heat wave that killed all these people was on average 6°C warmer than normal, but this number may be consistent with a 2°C global average temperature increase, because the calculation of the global average includes the hottest and the coolest spots on the planet. To give some idea of how hot central Europe was in the summer of 2003, it helps to think about how abnormal this three-month heat wave was.

By any estimate that I can find, this period was the hottest on record for central Europe, and it was five standard deviations warmer than the average summer temperature. "Standard deviation" is a term that expresses how far away from an average any given point or temperature is. For example, one standard deviation greater than the average would account for 68 percent of all temperatures that were ever recorded to be warmer than average; two standard deviations would account for 98 percent of all temperatures; 99.7 percent of all weather observations will fall within three standard deviations away from the average. The summer heat wave in Europe was five standard deviations—well outside the boundaries of any observable heat wave pattern in human history. If you think in terms of the chances of observing such a heat wave as a daily observation, the odds of a five standard deviation level condition is about one in 1,700,000. There should be thousands of years between such continental heat waves.

Crop losses across Europe in 2003 were in the billions of euros; by some estimates, the mountain glaciers in Switzerland lost 10 percent of their total mass during the three-month heat wave, and huge rock slides brought down the sides of mountains as a result of the melting of permafrost that holds rocks to the core of mountains (Beniston and Diaz, 2004; Lynas, 2008). The major rivers throughout Europe were at record low levels, because rainfall decreased to less than a quarter of its June normal. Insufficient water in the major rivers influences irrigation, electricity generation, and navigation, in effect throwing the circulation systems of countries into disarray. Some computer modeling that explores the impact of a 2°C increase in global temperatures on the United States suggests that the snowpack on western mountain ranges may be reduced by up to 70 percent, with consequences for navigation, irrigation, and electricity production in an already water-stretched western region (Leung et al., 2004). One version of the western United States in a hotter climate is comparable to the center of Europe in 2003. This is by no means a desirable outcome but still one to which we could adapt.

It is easy to dismiss the events in Europe 2003, Binghamton 2006 and 2011, and the East Coast of the United States in 2012 as events so rare that they should not influence our future expectations. The alternative, of course, is that they are harbingers of things to come—the new normal—if we do nothing to change our carbon consumption. One way to discern whether these are indeed rare events or the new normal is to think about what the "average" means, because at the core, our discount of future benefits is a function of how we see that average and how quickly we see it changing. So p and q, in this sense, are driven by how easily Europe 2003 can be dismissed as a rare one-off event. Figure 5.1 presents a graph of the average temperature on Earth at some particular point in time. This graph is completely hypothetical and depicts a set of normally distributed data. The hypothetical mean temperature is at the peak, and the variation in daily temperatures distribute away from that mean. Since we are really interested in examining that 2°C increase in temperature, I have represented the global average as zero. The standard deviations are the vertical dotted lines; the temperature distribution is what we refer to as a normal distribution. The curve shows the frequency at which the average temperature would increase or decrease by a certain amount.

In reading this graph, it is simple to see that the average is also where the "most common" temperature change would be found. It is, of course, reasonable to expect very small changes in the average global temperature on a daily or yearly basis. The first set of dashed lines to the right and left of the average represents one standard deviation from that average, such that 68 percent of any one annual increase might be found in this range. The second set of dashed lines represents two standard deviations encompassing roughly 98 percent of any observed annual increases. If the big climatic events are a result of changes that we would only observe in the tails of the temperature distribution,

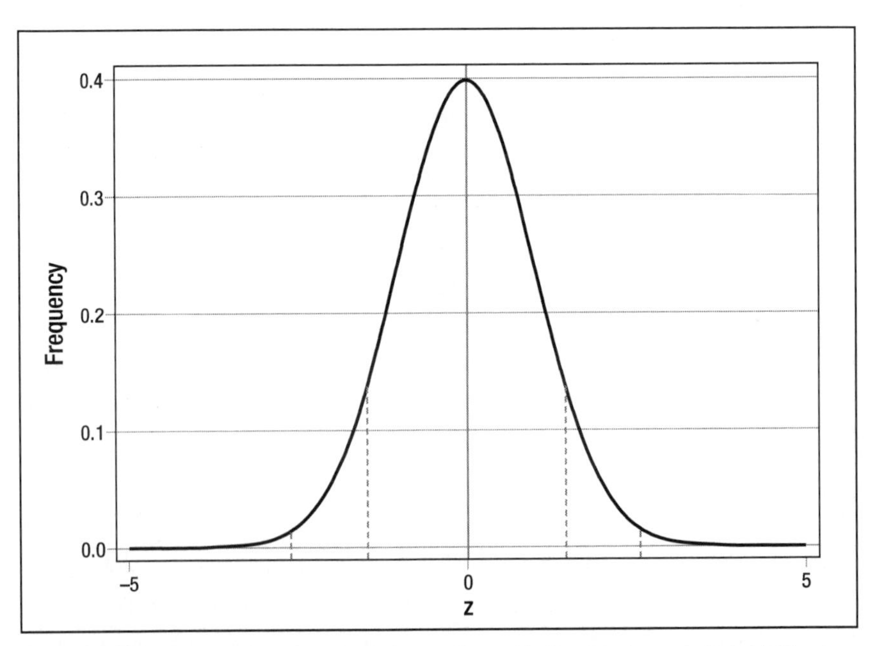

The X axis represents the hypothetical density of average temperatures when normalized to a mean of zero and a standard deviation of one. The Y axis represents the hypothetical frequency of observed temperatures.

Figure 5.1 Normal distribution of global temperatures.

then that is where we should look. Remember that the 2003 heat wave in Europe was five standard deviations outside the average, so in this graph, it would be to the right-hand side of the curve, and down almost where the curve touches the frame of the graph on the X axis. We should almost never see temperatures in that range, and indeed, the European heat wave was hotter than any similar period ever recorded. Even estimates of temperature through core sampling have to go back thousands of years to find Europe that hot. So if rare events—like 500-year floods that happen twice in six years, or a heat wave that kills tens of thousands, or superstorms that ravage coast lines outside hurricane belts—are associated with changes in our climate, we need to see what happens as the average temperature changes become systemic. That is, we want to see if events in the tail of a normal distribution begin to move closer to the normal as the average temperature increases.

Figure 5.2 has two other temperature distributions laid over the original graph from figure 5.1. The second and third curves in Figure 5.2 reflect a normal distribution with a new average temperature that is 2°C or 4°C warmer than the original. What can be readily seen is that temperatures occurring rarely under current conditions with a mean of zero become considerably more likely in a world warmer by 2°C, and fairly common when global temperatures are on average 4°C warmer. In fact, when the world is 2°C warmer than, say, 1950, the number of really hot days or heat-related events with a frequency in the ninety-fifth percentile would be the new average. To put this into context, the heat wave in the summer of 2003 that killed somewhere between 20,000 and 70,000 people in Europe was an extremely rare event under today's conditions, but we can expect to see these weather events much more frequently when the average temperature is 2°C warmer. A heat wave like the summer of 2003 would still be rare, but more on the order of once every 50 years rather than once every 4,000 years. Under conditions where the global climate is 4°C hotter, the 2003 European heat wave might be considered a cool summer rather than a traumatic event. If hurricanes that spawn in the Atlantic become fiercer and push further north as oceans get warmer, then 100- and 500-year flood maps in Binghamton, New York, will rapidly become obsolete, and superstorms along the East Coast of the United States, commonplace. The Binghamton floods, the European heat wave, and the storm that ravaged the East Coast occurred at temperatures only one-half of a degree warmer than they were 50 years ago and could legitimately be considered rare events today. Climate-change-related drought in the southwestern United States and torrential rains in the North have been projected for 30 years (Hansen, 1981).

We've got to make a bet on something, and the vast bulk of the scientific research has tied the devastating weather events of the past few years, the long-running drought in the western United States, excessive heat conditions, melting glaciers, and rising oceans to the small increase in global temperatures that we've experienced in the last half century. Since we are not even close to that 2°C increase yet, worse conditions are still to come.

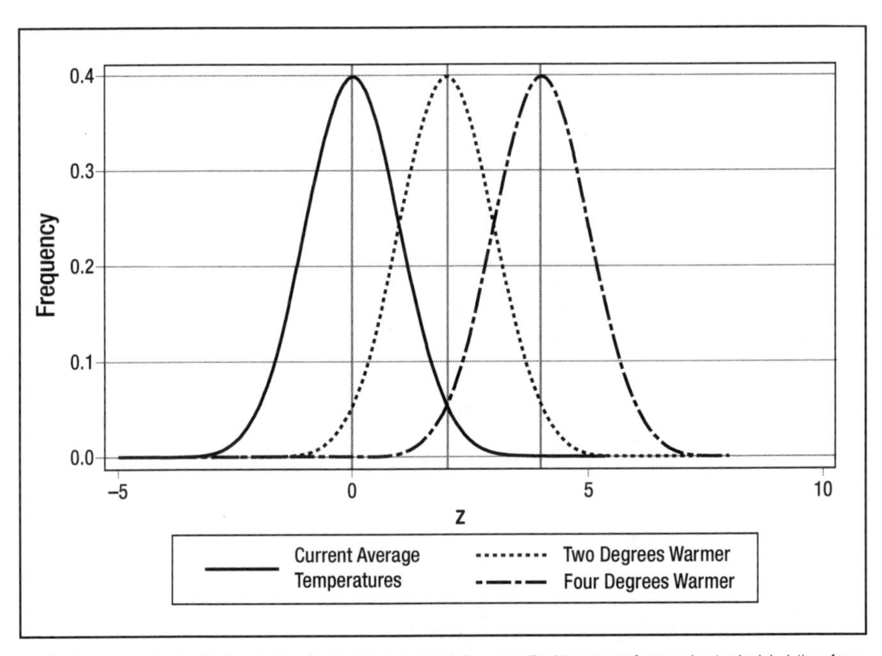

The X axis represents the hypothetical density of average temperatures when normalized to a mean of zero and a standard deviation of one. The Y axis represents the hypothetical frequency of observed temperatures at different normal states. The two-degree warmer mean position overlaps the current average distribution at the point of two standard deviations from the average, demonstrating that a global average that is 2°C warmer would be a rather rare temperature under current normal conditions. Vertical lines represent the average temperature at the respective means of the global average temperature.

Figure 5.2 Average at current conditions, 2°C and 4°C increase in global temperature.

From 1950 to 2002, two of Greenland's largest glaciers, Kangerdlugssuaq and Helheim, were stable. "Stability," in this sense, means they were neither charging forward nor retreating outside of a normal range. In 2002, the Heilheim glacier began shrinking. Over the next three years, it retreated 7 kilometers, or 4.2 miles. During the season of 2004–2005, the Kangerdlugssuaq glacier retreated 5 kilometers, or 3 miles. It was a bad year for the glaciers on Greenland. Not only did they retreat by a significant distance, but they got thinner by up to 80 meters (Howat et al., 2007). The Patagonia ice shelf in Argentina and Chile, the Antarctic ice shelves, and even Glacier National Park in Montana are showing the same pattern. Glacier National Park had 150 glaciers in 1850, and today, it has 26 (U.S. National Park Service). By the year 2020, the Park Service estimates that all the glaciers of Glacier National Park will be gone if we do nothing to reverse global warming. We're not even close to stabilizing at the 2°C increase that President Obama and other national leaders have committed to by the year 2020. So if the glacial ice on the planet is melting at this current rate when we are but a half of a degree warmer than we were 50 years ago, the rate of retreat of the world's glaciers as we approach that 2°C mark will be considerably more dramatic and destabilizing (National Climate Assessment, 2014).

By some estimates, the glaciers on Greenland shed 450 gigatons (Gt) of ice during the period from May 2004 through April 2006 (Velicogna and Wahr, 2006). To put the magnitude of gigaton quantities of ice into perspective, one gigaton of ice is equivalent to one cubic kilometer of water. To put this differently, a Gt would be a box one kilometer high, one kilometer wide, and one kilometer long, and each gigaton of ice would melt into a pool of water 12 Manhattan city blocks long, say, from Houston to 12th Street, four blocks going east to west, maybe 1st Avenue to 5th Avenue, and about as deep as two Empire State Buildings. In a two-year period, the Greenland glaciers shed 450 of those blocks of water in a world warmer by 0.5°C.

Although 450 cubic kilometers is a lot of water, relative to the 2.8 million cubic kilometers of ice on Greenland, it is only a drop in the bucket, so to speak. If the water released from the melting during the 2004–2006 period were spread over all the oceans, it would have resulted in a 1.3 millimeter increase in ocean levels—not enough to worry about in the very short term. But if the whole ice shelf were to melt, and this is a legitimate worry, it would lead to an increase in ocean levels of nearly 7 meters, or about 23 feet (Intergovernmental Panel on Climate Change, 2001). If ocean levels were to rise by 23 feet, hardly anything about the shape of the continental United States would be immediately recognizable. Some scientific reports suggest that if the planet warms by 1.6°C, we might reach a tipping point at which the meltiang of the Greenland ice is irreversible, and within 2,000 years, it could all be gone (Spott, 2012). But 2,000 years is a long enough time that it becomes easy to discount that future. The tipping point, however, might be within our lifetimes. Once reached, there might be nothing to do but wait and watch the waters rise. That tipping point, and not the distant melting of ice shelves, is what we should focus on as the future payoff that we might be inclined to discount.

It is a political question as to whether we can slow down the rate of increase in global temperatures sufficiently to max out at a level that is 2°C warmer than it was two decades ago. The task is really a matter of overcoming the collective action problem at the individual and political level so that we can begin to reduce our consumption of greenhouse-gas-emitting fossil fuels. Most scientific research concludes that there is little that can be done to prevent the planet warming by 2°C, but others see the possibility (Davenport, 2014a; Bianco et al., 2013). We see some glimpses of a world 2°C warmer when news of storms, droughts, melting glaciers, and summer-long heat waves fill the media. Humans are ingenious and can figure out how to adapt in a world that is only 2°C warmer, but two generations in the future will potentially see summer temperatures routinely above 100°F, and these days will seem to string together and last a lifetime. The rare climate events at that time will be those associated with a world 4°C hotter, a condition that is hard to observe because it is so rare given our current average temperature. The types of events and daily temperatures associated with this new normal would be in the small tail of the distribution in Figure 5.1, the type of day that you would see once in 500 years or less under today's conditions. In a 4°C world, these would be the new average or the most common occurrences in terms of weather. If we can't stop the increasing temperatures at 2°C, then a 4°C hotter world will become the norm.

The Worst of Possible Worlds

There is no clear picture of a world 4°C hotter than it was at the start of the 21st century. Although there have been times when the planet was significantly hotter than it is today, there were no humans then, so there is no directly recorded history of global temperatures. All the literature on the subject reflects estimates built on informed speculation. Scientists are able to develop some sense of what the world was like when it was 4°C hotter through ice coring techniques. Ice coring is one method for developing an understanding of temperatures before we had weather stations, satellites, and other methods for measuring weather patterns. Other means of estimating the temperature during periods before recorded history involve examining isotopes from fossils that can be effectively dated. If we can estimate global temperature during particular epochs, and the fossil records tell us what the geography was like and what types of animals were able to survive, we can develop a sense of what our planet might look like if we cross a threshold of a 4°C warmer climate.

Core samples from the Vostok research station in Antarctica suggest that the last time the earth was warmer than it is today was roughly 125,000 years ago. The temperatures during that period were around 4°C warmer than they are today, so it provides a useful metric to evaluate our future if we do nothing about climate change. It was roughly the time that anatomically human species began to appear on the planet (Stringer, 1994). Sea levels were about 20 feet higher than today, suggesting that ice melt and thermal expansion combined to reshape coastal boundaries (Weart, 2013). Groups of humans were beginning to migrate out of Africa 125,000 years ago, traveling across today's Sahara desert, which, at the time, was green with vegetation. The point should not be that the human species survived this period, but rather that those conditions ushered in some dramatic changes to the geography of the planet. Of course, some of us would survive a

4°C warmer world, but not all of us would. The population of human hunter gatherers at that time would have numbers in the low hundreds of thousands, not eight billion.

Records show that the world has been significantly hotter than it is today, particularly during the time of dinosaurs. We can also be reasonably certain that humans would not have found it a particularly pleasant environment. The planet has gone through numerous periods of warming and cooling over the past 4 billion years, including periods when there was no ice at all. In the Eocene epoch 40 million years ago, the earth was an iceless planet, and the atmosphere contained twice the carbon dioxide it does today. Primates did exist at that time, but they evolved into significantly smaller species, presumably to cope with the hotter climate that both preceded and followed. By most accounts, the earth was covered in forests; mountains were being formed, and ocean surface temperatures may have been as high as 95°F (Huber and Caballero, 2011). Humans did not have to contend with, nor have we evolved as a species able to cope with, such extreme temperatures.

We understand that a world without ice is going to change the makeup of the landmass that is inhabited. If Greenland's ice shelf melts, sea levels are estimated to rise by more than 20 feet, and a 4°C warmer climate is consistent with a time when Greenland was ice free. The National Oceanographic and Atmospheric Administration (NOAA) estimates that 50 percent of the U.S. population lives within 50 miles of a coastal area (NOAA, 2013), and nearly 85 percent of Australians live near the coast (Australian Bureau of Statistics, 2004). If sea levels rise by 20 feet, vast numbers of people would be displaced by these rising tides and a shrinking landmass. The land remaining will be mostly hotter and drier and less able to support the agricultural base required to sustain the current population. Models also project that some latitudes on some continents will have significantly more rainfall, to the point that too much rain may make subsistence difficult. In effect, under these increasingly hot conditions, the world will likely see mass migration into areas that are more secure from the rising oceans. This movement inland, however, will confront the challenges of increasingly arid landscapes and heat waves of the magnitude of Europe 2003, which would, by then, approach the new normal. People displaced from their homes and communities having to share dwindling resources will pose political and economic problems of some magnitude. Regardless of how climatologists describe an environment 4°C hotter, the politics may become completely debilitating for human interactions.

Climate, Resources, and Conflict on an Overheated Planet

One way to evaluate the risk of chaos resulting from competition over a shrinking pool of resources is to ask the organization charged with ensuring security. In 2011, the U.S. Department of Defense produced a report that evaluated the potential threat to international security from the climatic changes that we face. In the report entitled "Trends and Implications of Climate Change for National and International Security," the Defense Science Board concluded that climate change is likely to have its greatest impact on security through "its indirect effects on conflict and vulnerability" (Department of Defense, 2011, p. xi). Those charged with planning for future contingencies that pose a threat to U.S. and global security present a damning picture of the risk of conflict resulting from changes to our climate. Usually, military planners focus on political groups, arms flows, radicalism, and other such threats to security. Rain, drought, and encroaching shorelines are not in the typical purview of a military task force. This military task force, moreover, reports that the best estimates for what is often thought of as the most likely scenario (A1) is for a "4°C mean warming by 2070—or in the early 2060s if the carbon cycle feedback is stronger than predicted" (p. 16). Table 5.1 is from the task force report. It shows the potential tipping points for geological conditions and the temperature at which this tipping point may accelerate (p. 43). The transition time scale reflects the estimated speed that changes might take place. Arctic summer sea ice might be gone in 10 years.

The picture portrayed in this Defense Department analysis is striking for two reasons. First, it is remarkably consistent with IPCC reports and other scientific analyses of the impact of carbon emissions

Table 5.1 Tipping Elements in the Earth's Climate System

Tipping Element	*Feature of System, F (Direction of Change)*	*Control Parameter(s), ρ*	*Critical Value(s), †ρ_{crit}*	*Global Warming, †‡*	*Transition Timescale, †T*	*Key Impacts*
Arctic summer sea ice	Areal extent (–)	Local ΔT_{air}, ocean heat transport	Unidentified§	+0.5–2°C	≈10 yr (rapid)	Amplified warming, ecosystem change
Greenland ice sheet (GIS)	Ice volume (–)	Local ΔT_{air}	+≈3°C	+1–2°C	>300 yr (slow)	Sea level +2–7 m
West Antarctic ice sheet (WAIS)	Ice volume (–)	Local ΔT_{air}, or less ΔT_{ocean}	+≈5–8°C	+3–5°C	>300 yr (slow)	Sea level +5 m
Atlantic thermohaline circulation (THC)	Overturning (–)	Freshwater input to North Atlantic	+0.1–0.5 Sv	+3–5°C	≈100 yr (gradual)	Regional cooling, sea level, ITCZ shift
El Niño–Southern Oscillation (ESNSO)	Amplitude (+)	Thermocline depth, sharpness in EEP	Unidentified§	+3–6°C	≈100 yr (gradual)	Drought in Southeast Asia and elsewhere
Indian summer monsoon (ISM)	Rainfall (–)	Planetary albedo over India	0.5	N/A	≈1 yr (rapid)	Drought, decreased carrying capacity
Sahara/Sahel and West African monsoon (WAM)	Vegetation fraction (+)	Precipitation	100 mm/yr	+3–5°C	≈10 yr (rapid)	Increased carrying capacity
Amazon rainforest	Tree fraction (–)	Precipitation, dry season length	1,100 mm/yr	+3–4°C	≈50 yr (gradual)	Biodiversity loss, decreased rainfall
Boreal forest	Tree fraction (–)	Local ΔT_{air}	+≈7°C	+3–5°C	≈50 yr (gradual)	Biome switch

Table 5.1 Tipping Elements in the Earth's Climate System (continued)

Tipping Element	*Feature of System, F (Direction of Change)*	*Control Parameter(s), ρ*	*Critical Value(s),* †ρ_{crit}	*Global Warming,* †‡	*Transition Timescale,* †T	*Key Impacts*
Antarctic Bottom Water (AABW)*	Formation (–)	Precipitation–Evaporation	+100 mm/yr	Unclear¶	≈100 yr (gradual)	Ocean circulation, carbon storage
Tundra*	Tree fraction (+)	Growing degree days above zero	Missing‖	—	≈100 yr (gradual)	Amplified warming, biome switch
Permafrost*	Volume (–)	$\Delta T_{permafrost}$	Missing‖	—	<100 yr (gradual)	CH_4 and CO_2 release
Marine methane hydrates*	Hydrate volume (–)	$\Delta T_{sediment}$	Unidentified§	Unclear¶	10^3 to 10^5yr ($>T_E$)	Amplified global warming
Ocean anoxia*	Ocean anoxia (+)	Phosphorus input to ocean	+≈20 percent	Unclear¶	≈10^4 yr ($>T_E$)	Marine mass extinction
Arctic ozone*	Column depth (–)	Polar stratospheric cloud formation	195 K	Unclear¶	<1 yr (rapid)	Increased UV at surface

*Policy-relevant potential future tipping elements in the climate system and (below the empty line) candidates that we considered but failed to make the short list

† Numbers given are preliminary and derive from assessments by the experts at the workshop, aggregation of their opinions at the workshop, and review of the literature.

‡ Global mean temperature change above present (1980–1999) that corresponds to critical value of control, where this can be meaningfully related to global temperature.

§ Meaning theory, model results, or paleo-data suggest the existence of a critical threshold, but a numerical value is lacking in the literature.

¶ Meaning either a corresponding global warming range is not established or global warming is not the only or the dominant forcing.

|| Meaning no subcontinental scale critical threshold, could be identified, even though a local geographical threshold may exist.

Lenton, Timothy M. et al. 2008. "Tipping elements in the Earth's climate system," *Proceedings from the National Academy of Sciences*, vol. 105, no. 6: 1786–1793.

on future climate conditions. Second, it is written by the U.S. military, which has an abiding interest in ensuring our security and in getting it right. The report confirms the trend toward significantly warmer temperatures approaching the 4°C or higher mark. The main concern is not warmer temperatures and melting glaciers, which may be the simplest of our problems to resolve and develop adaption strategies. Armed conflicts that result from these dramatic changes may become a larger concern, and we can look to the social science community to shed light on whether or not these conflicts are related to pressures resulting from resource constraints.

There is some evidence that increased social conflict is a direct result of increases in temperatures and rainfall. In the prominent journal *Science*, a research team estimated the consequences from climate change in terms of personal and social violence, as well as armed conflict between groups (Hsiang et al., 2013). These results point to a 2.3 percent increase in interpersonal violence and a 13 percent increase in intergroup conflict for each standard deviation increase in temperature. A one standard deviation increase from the 1950–1980 baseline would approximate a 1–2°C increase in the global average temperature, so this is well within the range of where we are heading today. This study stands nearly alone in its precise predictions, but it does converge with the core implications of the Defense Department study (2011). But not all scholars develop such strong causal inferences about the relationship between climate and conflict, and in fact, many adopt a more cautious interpretation of the science behind climate-driven conflict (Exenberger and Pondorfer, 2013; Scheffran et al., 2012).

A recent issue of the *Journal of Peace Research* addressed the question of whether climate change is increasing the frequency of armed conflict (2012). The answer was inconclusive because of the uncertainty in the data. But a number of the studies do find direct positive links between climate variability and social conflict. Like the climate science itself, focusing on uncertainty could lead to the inference that there is no causal link between climate change and armed conflict, and therefore, minimize the cost. The climatic impacts we see today, however, are infrequent, dramatic, and abnormal. Most of the evidence in this recent issue supports the contention that as resources necessary to sustain stability dwindle, armed conflict becomes more likely. These results hold when the conflict involves rebellion and rioting against a state (Hendrix and Saleyan, 2012; Raliegh and Kniveton, 2012) or competition among pastoralists and herders (Butler and Gates, 2012). The results are not universal, however, with some studies failing to find a strong relationship between climate and conflict (Bergholt and Lujala, 2012; Gleditsch, 2012). The difficulty in predicting armed conflict based on contemporary weather data is in the ability to tease out the distinction between weather and climate. Short-term weather patterns are different from long-term climatic changes in the way that deviations from normal or mean temperatures are not the same as a new normal in the range of what was once extreme. Today, we can only look for extreme weather patterns within our normal climate conditions.

The data cannot implicate one large storm or one season of rough weather in the causal chain of armed conflict. At best, these relationships should be thought of in terms of triggers that push citizens to rebel. The pressures on the environment caused by long-term changes in the climate constrain or shrink the available resources to the point that people fight over them. There is a reasonable amount of evidence to demonstrate that people fight over access to resources, whether or not their availability is constrained by weather patterns (Gleditsch et al., 2006). So in our new 4°C warmer world, we need to think about resources, to whom they are available, how access changes with changing climate patterns, and how governments and people might respond to diminished resources. At a global level, humans really haven't had to confront these choices, yet, but at local levels, they have. But 100,000 years ago, the first humans to migrate into the Levant did not survive the changing climate that came with the Ice Age, and the remaining human populations were restricted to the African continent. What we do not know is whether this population died out because of the inability to cope with the natural elements, or whether they died out fighting over the dwindling resources available for them to survive.

Resources that people might fight over include many things. Two contemporary and familiar examples are water and diamonds. Conflicts over water and the boundaries created by waterways are common enough that they have a name: riparian conflicts. The Jordan River runs through Lebanon, Syria, Jordan, Israel, and the Palestinian territories. The primary source of the river is in Lebanon, in a Hezbollah-controlled region (Kaufman, 2009). The primary consumer of the water from the Jordan River is Israel. If Lebanon, Jordan, or Syria were to block the flow of the river into Israel, the result would be war. Israel has made this clear, and an international agreement codifies this relationship. Nils Petter Gleditsch and his colleagues (2006) make the empirical link to scarcity of water as the issue causing conflict rather than a river's role in demarcating a boundary. There are other such studies that make resources the incompatibility that generates armed conflict (Tir and Stinnett, 2012). "Blood diamonds" are an example illustrating the human propensity to fight over scarce and valuable things and suggest that as food, land, water, and housing become challenged by changing climate, they will become the things of value and the source of contention.

As with climate science, in order to think about the potential consequences from a significantly warmer planet, we have to project from what we know today toward possible future conditions. Conflict resulting from a scarcity of resources is one of those conditions that we need to consider, and we have to rely on the functional equivalent of ice cores, but once we reach that point of a 4°C warmer planet, it will be too late to let science tell us whether war and social chaos is more likely. We'll know it if and when we reach that point. And as with possible changes to our physical environment that could result from climate change, our social world is fraught with possible weak points that break under the strain of challenging weather, constrained resources, and migrating people.

Physical stability and sustenance, the ability to work, exchange mechanisms so we can individually specialize, and a form of order that dictates norms for acceptable rules and behaviors all hold the social world together. In an ice-free world with geographic boundaries inundated by the oceans, with droughts that make previously productive land inhospitable, and regions drenched in excess rainfall, people would have to find ways to reorient that social world. Today, we rely on trade and globalization of our economies. But when one group cannot produce enough to trade and yet doesn't have enough to survive, chaos may ensue. For example, the Defense Department (2011) report points to a 600,000-square-kilometer loss of cultivated land in Africa if the average global temperature rises 2.5°C. This is about 230,000 square miles—a tract of land 92 miles wide and 2,500 miles long, about the size of the fourth-largest state in the United States. Six hundred thousand square kilometers might seem like a huge tract of land, but it represents only 2 percent of the total landmass of the African continent. The area of the Sahara desert is about 30 percent of the total landmass of Africa and is largely incapable of sustained agriculture. That 2 percent reduction in arable land in the Defense Department report is on a continent of one billion people that is already experiencing food security challenges. The potential for social chaos on a planet that is 2.5°C warmer makes the concern in the report understandable (Brown and Funk, 2008).

If "beggar thy neighbor" policies—in which goods are produced and consumed internally as services become constrained—were to shut down trade, the prospect of interstate war between nuclear-armed countries may become a concern. The science does not allow for a certain conclusion. We could discount the future benefit of making changes now to prevent this menacing possibility and focus instead on the uncertainty of the science or our understanding of the relationship between resources and war to reassure ourselves that we might have it all wrong. But we would do that at our peril. We would do better to focus on p and q, our uncertainty and our willingness to discount the future. In an odd sort of a way, the viability of the human species may hang on such simple evaluations. It is not certain that there exist political institutions which can maintain order and peace when there are strong incentives to rebel. The United Nations as it is currently constituted is more effective when dealing with smaller and poorer countries, but when the struggles are in the United States or China or Russia, there is no guarantee that it will succeed. As a political scientist, my sense is that the whole system will slowly unravel and that democracy will prove incapable of constraining behavior that is tied to personal and group survival.

Planet Earth has experienced extreme climatic variations before, and survived. Whether humans can survive socially and politically is not clear. During prior hot periods in our planet's history, pythons, turtles, and other reptilians survived, maybe thrived, but older species of mammals declined. It is also evident from the record that the earth recovered from prior excessively hot periods by moving toward excessively cold ones, where glaciers covered large parts of the planet. Nature, it seems, has a pathway to these climatic changes, and it really doesn't need our help in moving between them.

The human species is incredibly resilient, as we have demonstrated through the long history of our evolution. However, one of the traits that we developed is a self-centered focus. Survival and fitness may have depended on it, but now, that same focus may be harming us. Individual altruism is no longer sufficient, nor is individual rational behavior. We need to pursue collective behavior to obtain the collective good of a secure planet, and we must do so at the national and international level. The results of doing anything less are not worth risking.

6

♦ ♦ ♦

MOVING FORWARD

PARIS AND BEYOND

Reducing CO_2 emissions enough to make a significant difference will require a huge collective effort. There are too many individuals who prefer the status quo, whose interests are threatened by change. The industrialist Koch brothers, for example, have waged a campaign against policies that would make solar energy more economically viable to the individual (*NYT*, 2014). The Koch brothers have interests in the coal business. They are against net metering, where individuals can sell excess electricity generated from household solar panels back to their electric company. The electric meter of such households spins backwards, creating a credit for the customer. The Koch brothers and others claim that this system threatens the coal industry because of its potential to reduce demand on coal-operated power stations. Federal and state subsidies provide incentives for homes or businesses to install solar panels, and the Koch brothers and others are trying to reduce or remove these incentives.

My point is not to suggest that any individual or group who finds a reason to impede progress toward mitigation of our deteriorating climate looks forward to the day when survivability of the species is at risk. I suspect that there is not one individual or group out there that wants to push the planet's ecosystem beyond the point of sustainability; it is just that immediate interests trump the long-term ones. It is easy to discount the future and question the anthropogenic causes behind climate change, particularly if your short-term interests benefit. The Koch brothers seem to epitomize this, and they have a lot of money to back up their private interests.

Incentivizing individual behavior is one pathway to reducing consumption, but not all of the steps for moving toward a sustainable planet require government incentives or convoluted regulations to force individual interests or behaviors. We could think about which methods to reduce carbon consumption are easy to achieve by simple behavioral changes, versus ones that need expensive technological efficiencies or national or international mandates.

In this concluding chapter, I will bring together some of the simple things we can do as individuals, while accepting that the problem is far from simple, and solutions go well beyond any individual changing their behavior. Collective responses are required at national and international levels, but if private interests impeded the ability of the Senate to ratify a treaty and Congress to pass legislation that moves the country toward sustainability goals, then the private interests of our elected representatives need to change that dynamic. Individuals can lead this change from the bottom up. This bottom-up demonstration effect, moreover, could accept a significant amount of free riders as long as the numbers of those who do participate are large.

The Politics from the Bottom Up

When the U.S. House of Representatives passed the American Clean Energy and Security Act (HR 2454) in June 2009, President Obama was a quick supporter. The narrowly passed bill called for CO_2 emissions to be cut by 18 percent from 2005 levels over the next decade. President Obama and many other international leaders thought on a bigger scale, with climate scientists encouraging them to do so. The target of the Copenhagen summit was a 50 percent reduction in CO_2

emissions, and for a host of reasons, including those described in Chapter 2, the summit failed to reach any meaningful and enforceable agreement. President Obama, however, did not give up on reducing U.S. emissions. He instead committed to administrative actions that would require a reduction of emissions by 17 percent from 2005 levels by the year 2020, a much more modest target than discussed at Copenhagen and still falling short of where we need to be, but it is a step in the right direction. His 2013 State of the Union address confirmed his commitment to putting the United States on the path to reducing consumption.

Nearly four years of the administrative capacity of the United States targeting CO_2 emissions have not put us on track to reach the modest goal of a reduction of 17 percent in our emissions by 2020. A recent report by the World Resource Institute (Bianco et al., 2013) suggests that if we maintain our current efforts, there will be little change in the magnitude of our emissions of CO_2. Under what WRI calls the "status quo scenario," we will not *increase* the rate that we discharge CO_2 into the atmosphere, but neither will we reduce it. Given growth in population and economic output, current efforts will only reduce our emissions relative to expectations of doing absolutely nothing. The WRI has articulated a "go-getter strategy," and to achieve the modest reduction of 17 percent involves, according to their estimation, reinventing how we consume—how we transport ourselves, heat our homes, cook our meals, and manufacture the products we need to maintain our standard of living. The go-getter strategy would put us on the path toward sustainable reductions consistent with the president's goal. The report recommends the phasing out of hydrofluorocarbons (HFCs) as one of the most easily achievable steps to the 17 percent goal. These HFCs replaced the CFCs that caused the hole in the ozone layer. Although their creation helped reverse ozone depletion, HFCs contribute to global warming and represent a significant part of the CO_2 emission problem. Today, the WRI estimates that HFCs account for 2 percent of the total greenhouse gas (GHG) emissions, but projects that by 2020, they will be the fourth most significant contributor to the total amount of GHGs we release into the atmosphere, and by 2035, the third most significant. Both levels are dwarfed by emissions from automobiles and power plants, but according to the WRI, HFC emissions represent one of the most fruitful areas in which reductions can be made. Being manufactured products, HFCs could be replaced, as was done with CFCs. We might think of this as low-hanging fruit that can be addressed at the level of industrial policy but not something that we as individuals get involved with.

According to the WRI, methane in the atmosphere is rising in direct proportion to the natural gas extracted from deep rock formations. "Fugitive" methane is a by-product of the hydrofracking process. Fracking exchanges consumption of carbon from coal or oil to carbon from natural gas, and though natural gas has a lighter carbon footprint, undesirable amounts of methane are released into the atmosphere by this process. If we continue to search for new ways to supply current and rising demand, new problems come into existence. The simplest and safest way to thwart this issue is to reduce the demand for coal, oil, and natural gas.

Fracking is debated furiously in New York and Pennsylvania. Pennsylvania allows open access to the hydrofracking method of extracting natural gas from shale rock formations. New York seems to be doing all it can to prevent the industry from moving into the state. The doubts and resistance come from local environmental consequences such as contaminated wells, chemical spills, and road destruction. This debate almost never turns to the impact of the hydrofracking process on climate change. Going there would raise the unpleasant question of overconsumption. To eliminate fracking from the political agenda, the need for carbon-based products would have to be eliminated. Much of the natural gas being extracted from the marcelus shale is being used to supplant coal in electric generation plants, and natural gas is a better alternative to coal from a CO_2 emission perspective. But if there is a demand for enormous volumes of natural gas, it will be supplied by processes that cause environmental damage in areas precious to someone. The debate about fracking should not really be one of "not in my backyard"—this is precisely the collective action problem where everyone wants the product but nobody wants to pay the cost for getting it. Nor should it be about where to get more gas to fulfill this ever-growing demand, but rather, how to reduce the need for a product that is causing so much environmental damage and political consternation. Reducing electricity consumption is the better alternative by far.

It is stunning that there is so little progress on reducing consumption in a time when the president in office is firmly committed to the reduction of CO_2 emissions and is willing to use the administrative capacity of the government to do so. The lack of progress is a pointed reminder of just how politically entrenched the dilemma is. When the issue was whales or wolves or ozone, even snail darters, there was almost no opposition to what was quite restrictive legislation (Yaffee, 1982). President Nixon, a Republican, introduced the core ideas behind the Endangered Species Act to a Democratically controlled Congress and the legislation passed almost without opposition. When it comes to CO_2 and our consumption of carbon, we can't even come close to adopting necessary changes, even when the president is on board and willing to use the government to circumvent congressional inaction. The national political climate is part of the problem and looms increasingly large as concentrations of CO_2 in the atmosphere continue to grow.

Since climate change, global warming, and their consequences are generally accepted facts, it is mostly the anthropogenic contribution in dispute. My reading of the scientific journals leaves me in little doubt that anthropogenic causes are at least a major contributor. There are some who question this relationship, and part of my argument is that motivational biases provide a rationale for them to do so. For example, the planet has just recently surpassed the 400 ppm level of CO_2 in the atmosphere, a level not seen on spaceship Earth in millions of years, and yet we continue to debate whether or not humans are driving the dramatic changes to our climate (Pew Research Center, 2013; Davenport, 2014a). We should be past this point of pedantic debates and instead focusing on what we can do to change the situation.

Private interests at the congressional level seem to ensure continued gridlock on climate change issues. Legislation supporting a sensible climate policy is a hard sell when a threat of a primary challenge against those who believe climate science data can be sufficient to derail a political campaign. The future discount from a stable climate is too high in some quarters. The pressure from above, the president, does not seem to work so understanding how bottom-up politics play out might save us from our own consumption patterns. Considering what we have learned about the likelihood of action driven by politicians themselves, democracy from below might be the strategy required for generating collective action.

On the face of it, individual-level efforts appear to be trivial when viewed in relationship to the magnitude of the overall problem of CO_2 emissions, but individual efforts can 1) contribute to significant reductions in CO_2 emissions, and 2) potentially drive politics at the national and international levels. Olson, after all, could have had it all wrong. A suboptimal outcome when the group is large might be sufficient to alter the preferences and incentives of the smaller groups enough that they find the future value of climate action to approximate the current value, that is, they cannot politically discount the future benefits from climate change mitigation policies. This still requires that there be benefits that outweigh costs from taking action (Olson, 1965; Sandler, 2004), but given the increased visibility of superstorms, huge tornadoes, and enduring droughts, the understanding of the benefits is getting more attention in the media. Private interests are tied to jobs and an established way of life. These can give way to shattered or flooded homes, parched farmlands, and an increased reliance on storm shelters for safety. We are seeing some of this in the media attention accorded the recent National Climate Assessment (NCA, 2014; Davenport, 2014b).

According to Olson's models, individual actions require individual altruism. I think we are better off believing the science and accepting that much of the CO_2 problem is man-made (IPCC, 2014), and then acting on it. To do otherwise is folly in the face of a serious challenge, and many of these "head in the sand" policies go horribly wrong (Tuchman, 1985). Once we accept the science, the questions need to focus on how willing we are to discount future benefits relative to contemporary costs and what we can do politically to change our plight. Most of us will discount future benefits when costs are high, but when current costs are marginal, there would be no need to do so. If we were given examples of action where the costs were minimal or imperceptible, and the resulting reductions in CO_2 significant, altruism would become irrelevant. In this sense, even Olson's collective action models would point to collective behavior in pursuit of a public good. If, moreover, politicians observed the public acting in ways that might appear to be against individual interests, they might focus on what climate scientists, hydrologists, and ecologists recommend and develop the political courage to put long-term interests over their short-term ones. This is asking a lot of our national politicians, who on average will not see the benefits of any climate mitigation actions; youth will. The average age of a U.S. House member is 57; in the Senate, the average age is 63 (111th Congress), so mitigation today will not dramatically change their world. If you are 20, it will change yours.

Very Low-Hanging Fruit

There are virtually cost-free steps individuals can take to minimize their own contribution to CO_2 emissions. For example, the WRI (Bianco, 2013) estimates that one-third of GHGs comes from electricity consumption, and the U.S. Energy Information Administration estimates that 18 percent of electricity consumption is attributable to residential use (EIA, 2014). This varies by state, but suggests that 6 percent of CO_2 emissions in the United States come from individual use within our homes (18 percent of 33 percent).

Roughly 7 percent of household electricity is used for lighting, just under half for appliances such as refrigerators, DVRs, and dishwashers. Heating, cooling, and other uses account for the remainder. To look at this simplistically, each household has to reduce electricity consumption by 50 percent to eliminate 3 percent of the annual CO_2 emissions from the United States. This approach very much requires individual initiative and would take a lot of individuals to participate before we could come close to achieving that goal. But if we did act, it would send a powerful message about our collective interest in a stable and sustainable environment. The fact that we would actually gain—in terms of a lower utility bill—is an added bonus.

There are several ways an individual might cut 50 percent from their current electricity usage. The U.S. government estimates that the average U.S. house used 11,000 kilowatt hours (kwh) of electricity in 2009 and that 60 percent of all refrigerators were either not energy star compliant or were more than nine years old. These are the inefficient units that could reduce by 30 percent the annual electricity usage from one appliance. With that comes a commensurate reduction in CO_2 emissions at the rate of two pounds of CO_2 for every kilowatt of electricity saved. About 75 percent of all washing machines in the United States are top-loading units, significantly less efficient than front loaders, and 80 percent of clothes dryers use electricity instead of natural gas (IEA, 2014). Exchanging old appliances for new ones is costly, particularly for the less well-off. National legislation could target these inequities.

The average HD DVR uses 63 watts of energy per hour, and this includes both when it is being used and when it is asleep (EUC, 2014). This average DVR uses nearly as much energy when sleeping as it does when someone is watching it (61 watts versus 62 watts), so there is no savings in just turning off the television. A little math will demonstrate how dramatic a small change can be in the use of an everyday product. If 50 percent of U.S. households have a DVR (roughly 50 million homes), the average DVR uses 1.5 kilowatt hours of electricity a day (63 watts × 24 hours), and it does this for a year, the annual usage of electricity amounts to 552 kilowatt hours per household. The annual usage by a single house multiplied over half of all houses in the United States amounts to a whopping 27.6 billion kilowatt hours of electricity in a single year because of the poor design of a single type of household appliance. DVR manufacturers think that we demand an instant-on DVR, and that we are willing to pay for that both in the increased cost of our electricity and in the consequences for our climate. A billion kilowatts is a gigawatt, and some of the larger coal-fired power plants produce gigawatts of energy. A medium-sized coal power plant produces roughly 500 megawatts, which is one-half of a gigawatt of energy each hour. Shutting off 50 million DVRs for 12 hours each day would reduce the emissions of CO_2 by the equivalent of taking offline one medium-sized coal plant for a half a day every year. At the rate of two pounds of CO_2 for each kilowatt of energy produced in a bituminous coal plant and 14 gigawatts of energy saved, we would reduce CO_2 emissions by something on the order of 28 million pounds, which translates into roughly 14,000 tons of CO_2 currently being emitted into the atmosphere. Fourteen thousand tons each year is but a fraction of the total amount of CO_2 we emit from coal-fired power plants, but it is an easy 14,000 tons. All it would take is attaching our DVR to a power strip on a timer.

Conspicuous consumption of electricity is so common, so normal, that we often look right past it. I was meeting a colleague for lunch at Brothers Bar and Grill in South Bend, Indiana. It was a bright, sunny summer day. As I waited out in front of the building, I noticed the exterior lights on the building were on. In all, there were nine lights on. Because the ambient sunlight was brighter than the lights, you could only see the lights by looking up at them. There were also four spotlights pointed at the Brothers sign that had little effect because of the bright light from the sun. I went into the restaurant and asked if they would turn off the lights that were only using electricity without making the building or the signage more visible. The hostess couldn't manage it, and the manager informed me that "corporate" required the lights on at all times. This is not a tremendous amount of electricity consumption, and the contribution of 13 lightbulbs to our problem of dealing with climate change is almost imperceptible. However, the amount of CO_2 emissions that result from those marginally useful lights is calculable with a few basic assumptions.

If I assume that each lightbulb used 100 watts and each spotlight used 150 watts, that one Brothers Bar and Grill was using 1.6 kilowatt hours of electricity each hour of the day for no observable purpose. There are 17 Brothers Bar and Grill restaurants in 10 cities around the country, and if we assume that each has the same number of lights, and all the local staff follow corporate policy, then for each hour of daylight, Brothers is consuming 27.2 kilowatt hours of electricity. And if I assume that there are six hours of daylight where the lighting is unnecessary to generate business, on a given day, Brothers Bar and Grill is wasting 163 kilowatt hours of electricity, over the course of a month 4,896 kilowatt hours of electricity, and over the year 58,752 kilowatt hours of electricity is wasted. This is just one restaurant chain acting on their corporate policy.

At 10 cents per kilowatt hour, the policy of leaving the lights on costs roughly $6,000 over the year. If the corporation calculates that leaving lights on during the daylight hours brings in more than $6,000 in business, from a corporate perspective, it makes sense. From a climate perspective, it appears to be abject waste, but even as waste, it is trivial to the overall magnitude of the problem we face. At a rate of two pounds of CO_2 per kilowatt hour of electricity, Brothers Bar and Grill is emitting 117,000 pounds of CO_2 into the atmosphere each year (EIA, 2014). But one small restaurant chain changing their policy of wasting 58,000 kilowatt hours that discharges 117,000 pounds of CO_2 will not put us on a path to sustainability. The problem dwarfs that simple fix. But this behavior is an example of how individuals as part of a collective can make a change. If every bar and grill restaurant, every McDonalds, and every Burger King leaves exterior lights on during the day, the implications are more consequential. I suspect that this corporate policy extends well beyond Brothers Bar and Grill.

So a vast amount of our carbon consumption problem falls into the category of wasteful, such that the everyday efforts could actually make a difference. Another example came from a late-morning run one weekend.

During a section of a class on climate change, I suggested that students who walk to campus count the number of porch lights left on during daylight hours. I did the same on my run one bright Saturday morning. I took a two-mile run in my subdivision, and I counted 38 lights that had been left on. Later that day, I drove two miles in the other direction. This time, I counted 41. By my small sample, on average 19 lights per mile were left on during hours when sunlight does the work of illumination. My students came in with slightly higher numbers of lights per mile. It became a simple calculation to determine how much CO_2 is emitted by leaving porch lights on during the day. Using an assumption of 40 watts per bulb, two million miles of residential roads, and 20 bulbs per mile, we consume 1.6 million kilowatt hours of electricity to power those porch lights for an hour. If they are left on for four hours before someone turns them off, they consume 6.4 million kilowatt hours of electricity in that day. The peak output of the Hoover Dam is 2.0 billion kilowatts, and the average-size coal power plant produces 500,000 kilowatts of electricity at peak performance (IEA, 2014). So at peak performance, it takes a dozen average coal plants one hour to power porch lights left on for four hours. The fix is easy: either diligence or a onetime investment in technology using a common light sensor.

The amount of CO_2 we emit during those four hours with unnecessary porch lights left on is roughly 6,400 tons. It is, like the restaurant lights, imperceptible relative to the total tonnage of CO_2 from coal-fired power plants, and the planet's ecosystem will not be stabilized just by turning off porch lights. It is a small amount when viewed from the aggregate. But the 6,400 extra tons of CO_2 those porch lights create is not helping solve our climate problems either.

Sixty-four hundred tons of CO_2 from porch lights, 58 tons from a change in corporate policy by the Brothers Bar and Grill chain, and 14,000 tons from realigning our relationship with the always-on DVR, however, add up to 41 million pounds of CO_2 left out of the atmosphere for one year as a result of simple behavioral changes. That is decidedly nontrivial.

Such examples abound, and they appear both trivial and profound. We waste considerable amounts of electricity in ways that, if changed, we would not even notice. Each kilowatt we waste translates directly into CO_2 emissions, and each one impedes our ability to meet reductions that would put us on the path to sustainability. This is low-hanging fruit that could send a very strong political message. The incentive to reduce waste has to come from one of three areas: 1) altruism, 2) an understanding of the benefits of reducing CO_2 on our planet, or 3) a financial incentive. It doesn't matter how a person gets there, but it is important that they do. Just shutting down the DVR for 12 hours a day would save the average household \$27.50 a year (275 kwh × \$.10 per kwh) and would easily offset the cost of a power strip and timer in the first year. Not all solutions are so easy, but the outcomes from the harder ones can be more effective.

Fruit a Bit Higher Up the Tree

Some fruit is higher up the tree, and picking it requires a greater investment; this is where individual behavior coupled with national legislation would be most effective. Take, for example, a comparison of two cars used for transportation to and from work, to the grocery store, and on vacation. If one of the cars gets 13 miles more to the gallon than the other car, and each owner drives 15,000 miles per year, the difference in fuel consumption and CO_2 emissions between the two is quite remarkable, particularly if you aggregate this across the 100 million of the 240 million cars on the road in the United States. If 40 percent of the vehicles in the United States are replaced with vehicles that get 13 miles per gallon (mpg) better fuel economy, the effect would be significant. To put this into context, the average fuel economy for the U.S. fleet in 2012 was 24.6 mpg (Ingram, 2013). This average includes cars, light trucks, minivans, and SUVs. A 13 mpg increase is below the current CAFE requirements, so well within the range of conversion to more efficient forms of transportation. And just to make clear that a 13 mpg increase is not asking a whole lot of the U.S. car fleet; both European and Japanese fleets already had an average fuel consumption in 2010 that was about 18 mpg better than the U.S. fleet of cars and light vehicles. In 2010, China and South Korea had significantly more efficient vehicles on the road than the United States did (C2ES, 2014) so projecting forward to a car that is 13 mpg more efficient is not a stretch. Moreover, fuel efficient cars do not require a change in lifestyle, as, for example, switching to public transportation would.

To demonstrate the impact, I'll work through the numbers. If I use 22 mpg for the least efficient car and 35 mpg for the more efficient one, the 22 mpg car will use 681 gallons of gas in the year; the more efficient car will use 428 gallons. The difference in a weekly fill-up is 5 gallons, which translates into about $17.50 less in fuels bills. That's not quite enough to get someone to buy a new car, because the payoff is too small and the future discount is too high. But if we think of this from the aggregate, that is, increasing 40 percent of the U.S. fleet of vehicles by 13 mpg, the impact is significant.

The U.S. government estimates emission rates of 16.68 pounds of CO_2 per gallon of gas with a 10 percent ethanol content (IEA, 2014). If we were to replace 100 million cars in the U.S. fleet with vehicles that got 13 mpg greater fuel efficiency, our annual CO_2 emissions would be reduced by over 200 million metric tons. Each new and more efficient car would use 253 fewer gallons of gas for the year, and at 16.68 pounds per gallon, the more efficient car would reduce CO_2 emissions by 4,224 pounds. In itself, this is probably not sufficient to motivate any one individual to invest in a new car. But if you aggregate this over 100 million cars, it generates an annual reduction of 14 percent of the total CO_2 emissions. The auto industry would love the idea of selling 100 million new cars, and in fact, trying to accomplish this feat would take at least six years at the annual production rate of 15 million cars per year. It would be a costly proposition at the individual level, but the return from an increase in jobs and economic growth would only add to the benefits accrued to our ecosystem. This would have to be motivated at the national level to effect the speed with which we change the fuel characteristics of our fleet of cars. Current, new CAFE standards have us moving squarely in that direction, but the market might not be the only way to get there. Incentivizing the purchase of more efficient vehicles—as was done in 2009—could be welcomed in the auto and supplier industries, and the costs recouped from the increase on income and sales taxes could help offset the costs of incentives.

A 13 mpg increase in fuel efficiency over 40 percent of the U.S. fleet would reduce CO_2 emissions by more than 200 million tons per year. In 2012, the total emissions of CO_2 from the U.S. fleet of cars and light trucks amounted to 1,551 million metric tons, 29 percent of our total energy-related emissions. Changing over our fleet of vehicles would reduce our total emissions by 14 percent from the previous year, and the reductions would be permanent in the sense that we would never go back to the older—and higher—consumption levels.

Household use of electricity varies quite considerably by state. The average house uses 900 kilowatt hours of electricity each month, with a standard deviation of about 200 kwh (IAE, 2014). This distribution suggests that 68 percent of the houses in the United States use between 700 and 1100 kilowatt hours of electricity each month. Lots of things differ across states and regions, including the price of electricity. In some instances, this is a function of the distribution costs of getting the electricity from a power generating plant into homes; in others, it reflects the means of production. Water, coal, wind, solar, and natural gas have different price structures. Our willingness to conserve on the use of electricity—to pay that individual cost for the collective good—is a function of the price we have to pay. Those who pay more for their electricity find ways to use less, and by using less, we emit less CO_2 into the atmosphere. One explanation for different patterns of usage is that different regions or states have different demands for heating or cooling. Possibly the use of electricity for cooling a house is greater than that for heating, or perhaps homes in one region are systematically larger than another. The price of electricity, however, does a good job of predicting the average monthly household usage. Figure 6.1 is a plot of the price per kilowatt of electricity by state and the number of kilowatt hours used on average in each state. The solid line is a regression fit of the data, reflecting the average relationship. If I exclude the cost of electricity in Hawaii, which is more than twice as expensive as the next closest state, a one cent per kilowatt hour increase in the price of electricity results in 63 fewer kilowatt hours used on average per household per month. And at two pounds of CO_2 per kilowatt hour, that one cent increase in the price of electricity reduces emissions by 126 pounds per month per household.

It might seem trivial to think that a small change in the price of electricity leads to significant conservation, but this is really what the collective action problem is all about. When it is cheap to waste, we tend to waste because it costs so little to do so. And if each individual discounts the future value of any conservation efforts, the small cost in electricity becomes imperceptible. Some states have programs to reduce consumption by incentivizing the adoption of more efficient means of heating, cooling, or cleaning clothes and dishes. Significant reductions in electricity consumption are also possible with simple behavioral changes, like the DVR, turning off porch lights during the day, and installing CFC or LED lightbulbs.

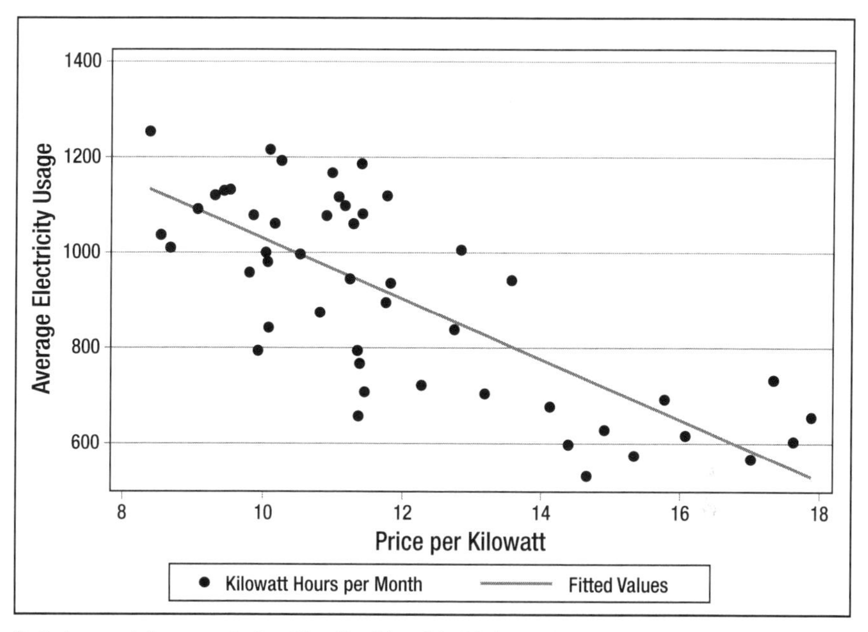

The X axis represents the average price (in cents) per kilowatt hour of electricity by state; the Y axis, the average monthly household consumption of electricity (in kilowatt hours). Graph excludes consumption and pricing in Hawaii. Data derived from the U.S. Energy Information Administration, 2014.

Figure 6.1 Average electricity usage relative to the price of electricity.

It might also come as no surprise that those households that tend to be the highest consumers of electricity come from states where coal and oil provide for a significant part of the local economy. Louisiana, Oklahoma, Texas, Kentucky, Tennessee, Mississippi, and Alabama are the highest household consumers of electricity, so formal restrictions on the burning of fossil fuels are going to be a hard sell. Each of these states is highly invested in the coal and oil industries, and states with a large number of employees in these industries are less likely to support climate mitigation policies (see Chapter 2). But at the individual level, it should be possible to bring household consumption in line with national averages, or even to target the consumption levels of some of the least-consuming states. For example, if you could get the average household in Louisiana to reduce its monthly electricity consumption from 1,200 kilowatt hours to 1,000 kwh, a 20 percent reduction from their current consumption levels but still 10 percent above the national average, those households would collectively reduce electricity use by 3.8 billion kilowatt hours over a year. At two pounds of CO_2 emitted per kilowatt hour, that would reduce the CO_2 emissions by more than 3.8 million tons. This is not low-hanging fruit, but it can be found in the DVR, temperature settings, and efficient lighting. The collective efforts of our neighbors in Louisiana would reduce our overall emissions by less than 1 percent of the total, but if the top 25 states targeted reductions on this order of magnitude, we would cut back 95 million tons of CO_2 each year, and these reductions would become the new status quo. This is not a herculean task and would have an immediate payback in terms of monthly electric bills.

All the low- or high-hanging fruit is not going to be sufficient to stabilize our climate if it is viewed solely from the perspective of an individual-level effort. The problem is much bigger, and our targeted reductions much more substantial than lightbulbs, DVRs, and new refrigerators can influence, and my point is not that we should point the way forward by seeking only the easiest solutions. We can't get there this way; solutions have to be much more systemic. My purpose for pointing to the easy steps is so that individuals can begin to see how local efforts can influence the types of policy initiatives required to comply with treaty obligations. Success on the climate change issue will require efforts on many levels and in many areas.

My argument has been a mix of increasing knowledge about causal processes, illuminating political dynamics that impede action, and increasing risk awareness of the consequences of doing nothing. O'Connor et al. (1999) demonstrate that people respond behaviorally based on knowledge and risk. As the perception of the risk of dramatic climatic events increases, so too do expected behavioral changes. An unfortunate implication of this risk-to-change dynamic is that we will only generate significant willingness to make behavioral changes as the point of no return approaches. Our response to ozone depletion made this clear. Those who understand the core relationships and already see an elevated risk from climate change need to put themselves at the cutting edge of behavioral changes. Santos and Pacheco (2011) argue that small groups are better able to evaluate risk and generate collective behavior, and importantly, coordination across groups can increase chances of success within groups.

If what we can achieve in the short term is at best a collection of individual efforts that together have a significant but not deterministic effect on climate stability, but in unison have the effect of coordinating actions that motivate higher level group behavior, then this might be the starting point. For those who understand the relationship between our carbon-based lifestyle and climate change and who accept that there is a significant risk of catastrophic consequences if we do nothing, then p and q in the model presented earlier approach the value of 1.0. We know from Olson (1965) and Sandler (2004) that if p and q are 1, action on the public good of climate stability reverts to a simple calculation that the benefits from action must be greater than the cost of taking it (EV = B–C). This is not an impossible situation to consider. Turning off a DVR for 12 hours a day and porch lights when they are not needed would have payback almost immediately, so for much of the low-hanging fruit, this should already present a positive expected value. The more costly steps—fuel-efficient cars, appliances, and insulation—will require incentives or restrictive legislation before compliance will be sufficient to offset significant amounts of CO_2. Bottom-up politics, disseminating knowledge, and a clearer understanding of the immediacy of the risk might move even this current congressional class to the point of cooperating rather than impeding efforts to stabilize our ecosystem. The most recent National Climate Assessment makes this clear (2014). The objective must be to get Congress to enact and the president to sign legislation that will entice or coerce behaviors across the broad swath of U.S. society—individuals, industry, and government—that will alter the ways, and the amounts, of carbon dioxide, we emit into the atmosphere.

Until Congress can act, larger steps that have significantly greater impacts on CO_2 emissions will have to come from administrative policy. Individuals can support these measures and use political pressure to compel our elected officials to do so as well. For example, the Supreme Court recently upheld the Obama administration's use of the EPA to impose new standards for emissions from newly built coal power plants (Davenport, 2014b). Current estimates are that each kilowatt of electricity produced from coal generates two pounds of CO_2, and under the standards recently articulated by the EPA, new coal-fired plants will have to cut that emission level nearly in half (Federal Register, 2014). Given that we generate nearly 1.5 billion tons of CO_2 through electricity generated by coal, this change will dwarf the impact that any individual can accomplish by turning off lights. The EPA's standards generated considerable resistance at the state and congressional level, and this resistance has to change. Individual behavior may send the message that collective behavior is in our interests (2013b).

The Requirement of International Cooperation

The problem of climate mitigation is not something unique to the United States, and in fact, most other countries have been more willing to sign and ratify international treaties than we have. And the problem of anthropogenically caused climate change is so big that even if the United States were to achieve its target of 17 percent reductions by 2020 or 50 percent by 2050, the emissions by the rest of the world would still be sufficient to increase climatic instability. Ultimately, the largest countries will have to make policies of sustained mitigation. China, Brazil, India, and other large countries know that they have to change consumption patterns, but it is hard for them as well. China seemed to be the target of much derision in the aftermath of the failed Copenhagen summit, because they were unwilling to forgo economic development in pursuit of a stable climate. This sounded a lot like the claims of some in the United States, that jobs will be sacrificed in pursuit of uncertain changes to our climate, and that jobs are immediate, whereas climate mitigation is long term.

China has a billion people and is trying to move hundreds of millions up the development ladder. To get there takes energy. But China has also taken steps to demonstrate it recognizes the problems at hand (Stensdal, 2012), and certainly during the Olympics, it became clear that emissions were a troubling aspect of the Chinese economy.

We can try to view this problem from the perspective of "we won't act until they do," or we can act in ways consistent with U.S. capabilities and bring the Chinese along. A stable climate is a public good, and any good we do is just as good for us as it is for them. That is, a coherent national response to climate change is simultaneously good for us and good for the rest of the world. We should accept that. China faces the same long-term versus short-term incentive structures that many in the United States face, though China's problems are exacerbated by numbers and the current level of economic development. India, as well as Brazil, might be in an even more difficult position. The solution is not to wait for them to catch up to us or Europe in mitigation policy but rather to lead the way, to make changes that reduce our carbon footprint and to do what we can—as individuals and as a society—to contribute to the stabilization of the planet. Anything else would appear to be political folly (Tuchman, 1985).

We know a lot about mitigation at local levels, and part of our collective problem is getting past the concerns at the national level so we can commit to international treaties. The utility companies that provide our electricity, often through coal-fired generation plants, will quite often provide subsidies for reductions in home consumption. They understand that aggregating the conservation efforts of many homes adds up to significant and measurable reductions in demand on their plants. A number of the larger cities around the United States have developed climate action plans in response to the unique climates in concentrated urban areas. The mayors of Los Angeles, New York, and Houston have each committed to planting one million trees in their respective cities because the trees help retain moisture, which, in turn, moderates the local climate (Stone, 2012). And many of our large industrial concerns have developed strategic plans to reduce their CO_2 emissions (Hoffman, 2007). There are recognized pathways to systemic reductions in CO_2 emissions from a broad spectrum of U.S. society, but individually, these will remain too meager until national-level political attention is brought to bear. All these individual efforts can serve as a model, but Congress has to take the next step. The problem of climate change is political.

Adaptation

Adaptation will be necessary whether we focus on mitigation or not. But there are two problems with focusing on adaptation as the primary strategy. The first is that adaptation to increasing pressures from our changing climate is a form of human social evolution. A critical question is whether the rate of change in pressures forced by our changing climate will outpace our ability to adapt. Today, we are able to put up higher levees, build homes on stilts and off the flood plains, or plan irrigation systems that can respond to increasing water scarcity. But we are only marginally capable of keeping up with this demand today, when the effects of climate change are reasonably muted. If the geophysical implications of climate change increase at the expected nonlinear rate, our ability to adapt at that pace will be called into question. If 20 years of climate science projections are correct, the pressures on the planet will increase sharply in the coming decades. Adaptation requires forethought and political expediency to distribute scarce resources to adaptation projects rather than military, social welfare, schools, or any number of competing claims on those same resources. In today's political environment, at least in the United States, that seems like a tall order. We also have to question how high we can build levees, how far from the flood plains we can move communities, and how effective our early warning systems are. The answers might not be consistent with the projected sea level rises, the expected future ocean temperatures, or the strength of megastorms. An adaptation-first strategy requires that we evolve socially at a pace and in areas consistent with changes in the ecosystem to which we are adapting. If social adaptation proceeds at a pace that is outstripped by geophysical processes, adaptation is but a short-term palliative targeted at a current generation. The political problems with this strategic approach are legion, and we know little about how to design and implement them effectively (Javeline, 2014).

The second problem with adaptation is that it is a response for the wealthy almost to the exclusion of the less well off. As it gets increasingly challenging to confront weather-related pressures on crops, resort towns, and flood zones, the United States will be able to cope through adaptation in the short term. We might not be able to keep pace over the long term, but current generations will see the immediate benefits. The countries in Africa that the U.S. Department of Defense (2011) estimates will lose 600,000 square kilometers of currently productive agricultural land will have a much more difficult time. This shifts the collective action problem radically from one of a global response to national or even local levels. Adaptation resources and strategies will make a big difference in who can respond and how, and wealthy countries, states, and towns will have more degrees of freedom than poorer ones. From this perspective, an adapt-first strategy is something of a stopgap, where the world community refuses to confront the global problem and instead devolves into problem solvers at increasingly smaller units of observation. If it is wealth that determines who adapts and who can't, who lives and who dies as a result of climate change pressures, we will have moved a long way from accepting climate stability as a public good to instead thinking about survivability and viability as private goods purchased by those communities who can.

As a species, humans have been remarkably adept at adapting to new environments, both physical and social (Wilson, 2002). To a large extent, this is the history of the human species, from the ability to walk upright, to hunting and cooking meat, to agricultural development, and ultimately to large-scale political and social communities. There should be no expectation that, as a species, humans will not continue to adapt to the environment. To me, climate change poses a more puzzling question of adaptation. Most of our adaptive strategies have been in pursuit of survival and fitness of the species in response to conditions nature presented. With climate change, it is largely humans who are generating the need for social adaptation, and to a significant degree, it is the wealthy human social communities who cause the problem and who have the immediate capability to adapt. Without the ability to adequately adapt to the demands of a changing climate, the poor suffer—and maybe perish—at the hands of the wealthy. To accept this condition requires that we not think of ourselves as a global community but rather as a collection of local ones that compete, and in that search for survival, the competition will be asymmetric until the wealthy can no longer adapt as rapidly as the changing climate demands.

Paris Climate Summit, 2015

National and international efforts since the failures at Copenhagen have made some progress toward putting in place the mechanisms for carbon reductions, but without the force of an international treaty that would compel action. President Obama has raised CAFE standards on U.S. cars and his administration has mandated reductions in CO_2 emissions from electric-power generation. The first generated little domestic opposition; the second considerable resistance. China has recognized the need to reduce emissions in spite of pressures to increase economic activity, and the European Union and other countries have put in place regulations that require reductions in CO_2 emissions. But all of this is without the enforceable commitments in international treaty. From this perspective these positive steps are all reversible without violating international constraints.

The Paris summit in 2015 is expected to change this. Following on a series of climate meetings in Durban, Doha, Warsaw, New York, and Lima, the Paris meetings will attempt to codify the limits that individual countries agree to target. The mechanism for assigning target levels of CO_2 emissions has changed—a lesson from the failure at Copenhagen—but the hope is that nationally-specified targets can be codified in treaty and then supported by domestic constituencies.

Not all are sanguine that useful national-level targets can be agreed upon, nor that implementation will follow easily from any agreement (Jacoby and Chen, 2014), but the meetings in New York, 2014, and the preparatory meeting in Mexico, 2015, are designed to get countries to put effective limits on the table. Paris will try to generate an agreement on those voluntary national commitments in CO_2 emissions, which, in turn, will be targeted at holding global temperature increases to 2°C. The modeling at MIT (2014) leave this outcome in doubt. Politics will be the key.

REFERENCES

American Clean Energy and Security Act, HR 2454. June 26, 2009. U.S. House of Representatives, 111th Congress, Washington, D.C.

Anderegg, William R. L., James W. Prall, Jocab Harold, and Stephen H. Schneider. "Expert Credibility in Climate Change." *Proceedings of the National Academy of Science* 107, no. 27 (2010): 12107–12109.

Australian Bureau of Statistics. "Yearbook Australia, 2004." http://www.abs.gov.au/Ausstats/abs@.nsf/Previousproducts/1301.0Feature%20Article32004 (accessed July 29, 2014).

Benedick, Richard Elliot. *Ozone Diplomacy: New Directions in Safeguarding the Planet.* Cambridge, MA: Harvard University Press, 1991.

Beniston, Martin, and Henry F. Diaz. "The 2003 Heat Wave as an Example of Summers in a Greenhouse Climate? Observations and Climate Model Simulations for Basel, Switzerland." *Global and Planetary Change* 44 (2004): 73–81.

Bergholt, Drago, and Päivi Lujala. "Climate-Related Natural Disasters, Economic Growth and Armed Civil Conflict." *Journal of Peace Research* 49, no. 1 (2012): 147–162.

Bernauer, Thomas. "Climate Change Politics." *Annual Review of Political Science* 16 (2013): 421–448.

Bianco, Nicholas M., Franz T. Litz, Kristin Igusky Meek, and Rebecca Gasper. "Can the US Get There From Here? Using Existing Federal Laws and State Action to Reduce Greenhouse Gas Emissions." Washington, D.C.: World Resources Institute, 2013.

Broder, John. "House Republicans Take EPA Chief to Task." *New York Times*, February 9, 2011.

———. "Both Romney and Obama Avoid Talk of Climate Change." *New York Times*, October 25, 2012.

Brown, David E. *The Wolf in the Southwest: The Making of an Endangered Species.* Tucson: University of Arizona Press, 1983.

Brown, Molly E., and Chris C. Funk. "Food Security under Climate Change." *Science* 219 (2008): 580–581.

Brunnee, Jutta. *Acid Rain and the Ozone Layer Depletion: International Law and Regulation.* Dobbs Ferry, NY: Transnational Publishers, 1988.

Butler, Christopher K., and Scott Gates. "African Range Wars: Climate, Conflict, and Property Rights." *Journal of Peace Research* 49, no.1 (2012): 23–34.

Calarne, Cinnamon Pinon. "Saving the Whales in the New Millennium: International Institutions, Recent Developments and the Future of International Whaling Policies." *Virginia Environmental Law Journal* 24, no. 1 (2005): 1–48.

Center for Climate and Energy Solutions (C2ES). "Comparison of Actual and Projected Fuel Economy for New Passenger Vehicles." 2014. http://www.c2es.org/federal/executive/vehicle-standards/fuel-economy-comparison (accessed April 29, 2014).

Chong, Dennis. *Rational Lives: Norms and Values in Politics and Society*. Chicago: University of Chicago Press, 2000.

———. *Collective Action and the Civil Rights Movement*. Chicago: University of Chicago Press, 1991.

Congressional Budget Office. "Agricultural Act of 2014." Washington, D.C., 2014.

Davenport, Coral. "Political Rifts Slow U.S. Effort on Climate Laws." *New York Times*, April 15, 2014. http://www.nytimes.com/2014/04/15/us/politics/political-rifts-slow-us-effort-on-climate-laws.html?_r=0 (accessed April 15, 2014).

———. "Justices Back Rule Limiting Coal Pollution: Victory for EPA and States on the East Coast." *New York Times*, April 30, 2014. http://www.nytimes.com/2014/04/30/us/politics/supreme-court-backs-epa-coal-pollution-rules.html?_r=0 (accessed July 29, 2014).

Department of Defense. "Report of the Defense Science Board Task Force on Trends and Implications of Climate Change for National and International Security." Washington, D.C.: Office of the Under Secretary of Defense for Acquisition, Technology, and Logistics, 2011.

DeSombre, Elizabeth. *The Global Environment and World Politics*. New York: Continuum, 2007.

Dietz, Thomas, Elinor Ostrom, and Paul Stern. "The Struggle to Govern the Commons." *Science* 302 (2003): 1907–1912.

Dimitrov, Radoslav S. "Inside Copenhagen: The State of Climate Governance." *Global Environmental Politics* 10, no. 2 (2010): 18–24.

Dolin, Eric Jay. *Leviathan: The History of Whaling in America*. New York: W. W. Norton and Company, 2007.

Dow, Kirstin, and Thomas E. Downing. *The Atlas of Climate Change: Mapping the World's Greatest Challenge*. Berkeley: University of California Press, 2007.

EIA (U.S. Energy Information Administration). "How Much Carbon Dioxide is Produced per Kilowatt Hour When Generating Electricity with Fossil Fuels?" 2014. http://www.eia.gov/tools/faqs/faq.cfm?id=74&t=11 (accessed April 20, 2014).

EPA. "Future Climate Change." 2013a. http://www.epa.gov/climatechange/science/future.html#Temperature (accessed, February 27, 2013).

———. "EPA Fact Sheet: Reducing Carbon Pollution for Power Plants." 2013b.

EUC (The Energy Use Calculator). "Electricity Usage of a DVR." 2014. http://energyusecalculator.com/electricity_dvr.htm (accessed May 1, 2014).

Exenberger, Andreas, and Andreas Pondorfer. "Climate Change and the Risk of Mass Violence: Africa in the 21st Century." *Peace Economics, Peace Science, and Public Policy* 19, no. 3 (2013): 381–392.

Federal Register: The Daily Journal of the U.S. Government. "Standards of Performance for Greenhouse Gas Emissions from New Stationary Sources: Electric Utility Generating Units." 2014. https://www.federalregister.gov/articles/2014/01/08/2013-28668/standards-of-performance-for-greenhouse-gas-emissions-from-new-stationary-sources-electric-utility (accessed April 30, 2014).

Fleming, James Rodger. *Historical Perspectives on Climate Change*. Oxford: Oxford University Press, 1998.

France Diplomatie. 2014. "Issues and Reasons Behind the French Offer to Host the 21st Conference of the Parties on Climate Change 2015," http://www.diplomatie.gouv.fr/en/french-foreign-policy-1/sustainable-development-1097/21st-conference-of-the-parties-on/article/issues-and-reasons-behind-the (accessed October 19, 2014).

Francis, Daniel. "Whaling," *The Canadian Encyclopedia*. The Historica-Dominion Institute, 2012. http://www.thecanadianencyclopedia.ca/en/article/whaling/. (accessed December 18, 2012).

Friedheim, Robert L., ed. *Toward a Sustainable Whaling Regime*. Seattle: University of Washington Press, 2001.

Gartzke, Erik, and J. Mark Wrighton. "Thinking Globally or Acting Locally? Determinants of the GATT Vote in Congress." *Legislative Studies Quarterly* 23, no. 1 (1998): 33–55.

Gleditsch, Nils Petter. "Whither the Weather? Climate Change and Conflict," *Journal of Peace Research* 49, no. 1 (2012): 3–10.

Gleditsch, Nils Petter, Kathryn Furlong, Håvard Hegre, Bethany Lacina, and Taylor Owen. "Conflicts over Shared Rivers: Resource Scarcity or Fuzzy Boundaries?" *Political Geography* 25, no. 4 (2006): 361–382.

Gray, Louise. "Copenhagen Climate Conference: US Says China Must Make Cuts." *The Telegraph*, December 9, 2009. http://www.telegraph.co.uk/earth/copenhagen-climate-change-confe/6772358/Copenhagen-climate-conference-US-says-China-must-make-cuts.html (accessed July 29, 2014).

Hamlet, A. F. "Assessing Water Resources Adaptive Capacity to Climate Change Impacts in the Pacific Northwest Region of North America." *Hydrology and Earth System Sciences* 15 (2011): 1427–1443.

Hansen, J., D. Johnson, A. Lacis, S. Lebedeff, P. Lee, D. Rind, and G. Russell. "Climate Impact of Increasing Atmospheric Carbon Dioxide." *Science* 213, no. 4511 (1981): 957–966.

Hanson, J., Makiko Sato, and Reto Ruedy. "Perception of Climate Change." *Proceedings of the National Academy of Sciences* 109, no. 37 (2012). www.pnas.org/cgi/doi/10.1073/pnas.1205276109.

Harrison, Kathryn, and Lisa McIntosh Sundstrom. "The Comparative Politics of Climate Change." *Global Environmental Politics* 7, no. 4 (2007): 1–18.

Hendrix, Cullen S., and Idean Salehyan. "Climate Change, Rainfall, and Social Conflict in Africa." *Journal of Peace Research* 49 (2012): 135–150.

Hoffman, Andrew. *Carbon Strategies: How Leading Companies Are Reducing Their Climate Change Footprint*. Ann Arbor: University of Michigan Press, 2007.

Howat, Ian M., Ian R. Joughin, and Ted A. Scambos. "Rapid Changes in Ice Discharge from Greenland Outlet Glaciers." *Science Express*, February 8, 2007: 1–10.

Hsiang, Solomon M., Marshall Burke, and Edward Miguel. "Quantifying the Influence of Climate on Human Conflict." *Science* 341 (September 13, 2013).

Huber, M., and R. Caballero. "The Early Eocene Equable Climate Problem Revisited." *Climate of the Past Discussions* 6 (2011): 241–304.

Ingram, Anthony. "Average Fuel Economy of US Cars Reaches an All-Time High." *Christian Science Monitor*. April 6, 2013. http://www.csmonitor.com/Business/In-Gear/2013

/0406/Average-fuel-economy-of-US-cars-reaches-an-all-time-high (accessed April 25, 2014).

Intergovernmental Panel on Climate Change (IPCC). "Climate Change 2014: Impacts, Adaptation, and Vulnerability." http://www.ipcc.ch/report/ar5/wg2/. Draft released March, 2014 (accessed July 28, 2014).

Jacoby, Henry D., and Y-H Henry Chen. 2014. "Expectations for a New Climate Agreement," MIT Joint Program on the Science and Policy of Climate Change, report #264, Cambridge: Massachusetts Institute of Technology.

Javeline, Debra. "The Most Important Topic Political Scientists Are Not Studying: Adapting to Climate Change." *Perspectives on Politics* 12 (2014): 420–434.

Kaufman, Asher. "Let Sleeping Dogs Lie: On Ghajar and Other Anomalies in the Syria-Lebanon-Israel Tri-Border Region." *The Middle East Journal* 63, no. 4 (Autumn 2009): 539–560.

Kauffman, Matthew J., Jedediah F. Brodie, and Erik S. Jules. "Are Wolves Saving Yellowstone's Aspen? A Landscape-Level Test of a Behaviorally Mediated Trophic Cascade." *Ecology* 91, no. 9 (2010): 2742–2755.

Laundre, John W., Lucina Hernandez, and Kelly B. Altendorf. "Wolves, Elk, and Bison: Reestabliishing the 'Landscape of Fear' in Yellowstone National Park, USA." *Canadian Journal of Zoology* 79 (2001): 1401–1409.

Lenton, Timothy M., Hermann Held, Elmar Kriegler, Jim W. Hall, Wolfgang Lucht, Stefan Rahmstorf, and Hans Joachim Schellnhuber. "Tipping Elements in the Earth's Climate System." *Proceedings from the National Academy of Sciences* 105, no. 6 (2008): 1786–1793.

Leung, L. Ruby, Yun Qian, Xindi Bian, Warren M. Washington, Jongil Han, and John O. Roads. "Mid-Century Ensemble Regional Climate Change Scenarios for the Western United States." *Climate Change* 62 (2004): 75–114.

Lichbach, Mark I. "Rethinking Rationality and Rebellion Theories of Collective Action and Problems of Collective Dissent." *Rationality and Society* 6 (January 1994): 8–39.

Lobell, David B., Marshall B. Burke, Claudia Tebaldi, Michael D. Mastrandrea, Walter P. Falcon, and Rosamond L. Naylor. "Prioritizing Climate Change Adaptation Needs for Food Security in 2030." *Science* 319 (2008): 607–610.

Loomis, John B., and Douglas S. White. "Economic Benefits of Rare and Endangered Species: Summary and Meta-Analysis." *Ecological Economics* 18 (1996): 197–206.

Lynas, Mark. *Six Degrees: Our Future on a Hotter Planet.* Washington, D.C.: National Geographic, 2008.

Maxwell, James, and Forrest Briscoe. "There's Money in the Air: The CFC Ban and DuPont's Regulatory Strategy." *Business Strategy and the Environment* 6 (1997): 176–286.

McMichael, Anthony J. *Planetary Overload: Global Environmental Change and the Health of the Human Species.* New York: Cambridge University Press, 1993.

McWhirter, Sheri. "Top Predators Descend: Lower Michigan Wolves Sighted; DNR Seeks Help in Survey." *The Leader and the Kalkaskian*, February 23, 2011. http://morningstarpublishing.com/articles/2011/02/23/leader_and_kalkaskian/news/doc4d652a1244e0c779390693.txt (accessed January 3, 2013).

Mech, David L. "The Challenge and Opportunity of Recovering Wolf Populations." *Conservation Biology* 9, no. 2 (1995): 1–9.

Michaels, Patrick J. "Science or Political Science: An Assessment of the US National Assessment of the Potential Consequences of Climate Variability and Change." In *Politicizing Science: The Alchemy of Policymaking*, edited by Michael Gough. Stanford, CA: Hoover Institution Press, 2003.

Michaels, Patrick M., and Robert C. Balling Jr. *Climate of Extremes: Global Warming Science They Don't Want You to Know*. Washington, D.C.: Cato Institute, 2009.

Minnesota Department of Natural Resources. "Rare Species Guide: *Canis lupus*, Linnaeus, the Grey Wolf." 2012. http://www.dnr.state.mn.us/rsg/profile.html?action=elementDetail&selectedElement=AMAJA01030 (accessed May 1, 2014).

Molina, J. Mario, and F. S. Rowland. "Stratospheric Sink for Chlorofluoromethanes: Chlorine Atomic-Atalysed Destruction of Ozone." *Nature* 249 (June 29, 1974): 810–812.

Motavalli, Jim. "With a Mileage Deal Near, the Auto Lobby Pulls Its Anti-CAFE Ads." *CBS News Money Watch*, 2011. http://www.cbsnews.com/news/with-a-mileage-deal-near-the-auto-lobby-pulls-its-anti-cafe-ads/ (accessed March 21, 2014).

National Aeronautics and Space Administration. "Global Climate Change: Vital Signs of the Planet." 2014. http://climate.nasa.gov/evidence (accessed May 12, 2014).

National Climate Assessment. *Climate Change Impacts in the United States*. Washington, D.C.: National Printing Office, 2014.

New York Times. "The Koch Attach on Solar Energy" (editorial), April 26, 2014. http://www.nytimes.com/2014/04/27/opinion/sunday/the-koch-attack-on-solar-energy.html?_r=0 (accessed April 28, 2014)

Nijssen, Bart, Greg M. O'Donnell, Alan F. Hamlet, and Dennis P. Lettenmair. "Hydrologic Sensitivity of Global Rivers to Climate Change." *Climate Change* 50 (2001): 143–175.

NOAA (National Oceanographic and Atmospheric Administration). "What Percentage of the American Population Lives near the Coast?" National Oceanographic and Atmospheric Administration Ocean Facts, 2013. http://oceanservice.noaa.gov/facts/population.html (accessed April 30, 2014).

O'Connor, Robert E., Richard J. Bord, and Ann Fisher. "Risk Perceptions, General Environmental Beliefs, and Willingness to Address Climate Change." *Risk Analysis* 19, no. 3 (1999): 461–471.

Oliver, Pamela. "Source Rewards and Punishments as Selective Incentives for Collective Action: Theoretical Investigations." *The American Journal of Sociology* 85, no. 6 (1980): 1356–1375.

Olivier, Jos G. J., Greet Janssens-Maenhout, Jeroen A. H. W. Peters, and Julian Wilson. "Long-Term Trend in Global CO_2 Emissions, 2011 Report." The Hague: Netherland Environmental Assessment Agency, 2011.

Olson, Mancur. *The Logic of Collective Action: Public Goods and the Theory of Groups*. Cambridge, MA: Harvard University Press, 1965.

Orr, James C., Victoria J. Fabry, Olivier Aumont, Laurent Bopp, Scott C. Doney, Richard A. Feely, Anand Gnanadesikan, Nicolas Gruber, Akio Ishida, Fortunat Joos, Rober M. Key, Keith Linddsay, Ernst Maier-Reimer, Richard Matear, Patrick Monfray, Anne

Mouchet, Raymond G. Jajjar, Gian-Kasper Plattner, Keith B. Rodgers, Christopher L. Sabine, Jorge L. Sarmiento, Reiner Schlitzer, Richard D. Slater, Ian J. Totterdell, Marie-France Weirig, Yasuhiro Yamanaka, and Andrew Yool. "Anthropogenic Ocean Acidification over the Twenty-First Century and Its Impact on Calcifying Organisms." *Nature* 437, no. 29 (2005): 681–686.

Ostrom, Elinor. "Coping with the Tragedies of the Commons." *Annual Review of Political Science* 2 (1999): 493–535.

Peterson, Shannon. "Congress and Charismatic Megafauna: A Legislative History of the Endangered Species Act." *Environmental Law* vol. 29 (1999): 463–492.

Pew Research Center. "Climate Change: Key Data Points from Pew Research," January 27, 2014, Washington, D.C. http://www.pewresearch.org/key-data-points/climate-change-key-data-points-from-pew-research/ (accessed July 29, 2014).

Plait, Phil. "The Earth is Warming Faster Now than It Has in 11,000 Years." Slate.com, 2013. http://www.slate.com/blogs/bad_astronomy/2013/03/13/global_warming_new_study_shows_warming_is_faster_than_it_has_been_in_11.html (accessed March 13, 2013).

Putnam, Robert D. "Diplomacy and Domestic Politics: The Logic of Two Level Games." *International Organization* 42, no. 3 (1988): 427–460.

Rahmstorf, Stefan, Anny Cazenave, John A. Chruch, James E. Hansen, Ralph F. Kelling, David E. Parker, and Richard C. J. Somerville. "Recent Climate Observations Compared to Projections." *Science* 316, no. 4 (May 2007): 709.

Raleigh, Clionadh, Lisa Jordan, and Idean Salehyan. "Assessing the Impact of Climate Change on Migration and Conflict." Social Development Department, World Bank, 2008.

Raleigh, Clionadh, and Dominic Kniveton. "Come Rain or Shine: An Analysis of Conflict and Climate Variability in East Africa." *Journal of Peace Research* 49, no. 1 (2012): 51–64.

Robine J. M., S. L. Cheung, S. Le Roy, H. Van Oyen, C. Griffiths, J. P. Michel, and F. R. Herrmann. "Death Toll Exceeded 70,000 in Europe during the Summer of 2003." *Comptes Rendus Biologies* 331, no. 2 (2008): 171–178.

Roosevelt, Margot. "Critics' Review Unexpectedly Supports Scientific Consensus on Global Warming." *Los Angeles Times*, April 4, 2011.

Rosenthal, Elizabeth. "Atop TV Sets, a Power Drain that Runs Nonstop." *New York Times*, June 25, 2011.

Ross, W. Gillies. *An Arctic Whaling Diary*. Toronto: University of Toronto Press, 1984.

Rypdal, Kristin, Terje Berntsen, Jan S. Fuglestvedt, Kristin Aunan, Asbjorn Torvanger, Frode Stordal, Jozef M. Pacyna, and Lynn P. Nygaard. "Tropospheric Ozone and Aerosols in Climate Agreements: Scientific and Political Challenges." *Environmental Science and Policy* 8 (2004): 29–43.

Sagan, Carl. *Cosmos*. New York: Random House, 1980.

Sagan, Scott D. "More Will Be Worse." In *The Spread of Nuclear Weapons: A Debate Renewed*, by Scott D. Sagan and Kenneth N. Waltz. New York: Norton, 2003.

Sandler, Todd. *Global Collective Action*. New York: Cambridge University Press, 2004.

Santos, Francisco C., and Jorge M. Pacheco. "Risk of Collective Failure provides an Escape from the Tragedy of the Commons." *Proceedings of the National Academy of Sciences* 108, no. 26 (2011): 10421–10425.

Schanning, Kevin. "Human Dimensions: Public Opinion Research Concerning Wolves in the Great Lakes States of Michigan, Minnesota, and Wisconsin." In *Recovery of Gray Wolves in the Great Lakes Region of the United States*, edited by Adrian P. Wydeven, Timothy R. VanDeelen, and Edward J. Heske. New York: Springer Press, 2009.

Schar, Christoph, Pier Luigi Vidale, Daniel Luthi, Christoph Frei, Chistina Haberli, Mark A. Liniger, and Christof Appenzeller. "The Role of Increasing Temperature Variability in European Summer Heatwaves." *Nature* (2004). doi:10.1038/nature02300.

Schar, Christoph, and Gerd Jendritzky. "Hot News from Summer 2003." *Nature* 432, no. 2 (December 2004): 559–560.

Scheffran, Jurgen, Miahcel Brzoska, Jasmin Kominek, P. Michael Link, and Janpeter Schilling. "Climate Change and Violent Conflict." *Science* 336, no. 18 (May 2012): 869–871.

Schneider, Stephen H., and Kristin Kuntz-Duriseti. "Uncertainty and Climate Change Policy." In *Climate Change Policy: A Survey*, edited by S. H. Schneider, A. Rosencranz, and J. O. Niles. Washington, D.C.: Island Press, 2002.

SEED (Schlumberger Excellence in Educational Development). "Global Climate Change and Energy Temperature Change History." 2013. http://www.planetseed.com/relatedarticle/temperature-change-history (accessed May 15, 2013).

Simmons, Beth A., and Daniel J. Hopkins. "The Constraining Power of International Treaties: Theory and Methods." *American Political Science Review* 99, no.4 (2005): 623–631.

Singer, S. Fred, and Dennis T. Avery. *Unstoppable Global Warming: Every 1500 Years*. New York: Rowman and Littlefield, 2007.

Slaper, Harry, Guus J. M. Velders, and Jan Matthijsen. "Ozone Depletion and Skin Cancer Incidence: A Source Risk Approach." *Journal of Hazardous Materials* 61 (1998): 77–84.

Solomon, S., D. Qin, M. Manning, Z. Chen, M. Marquis, K. B. Averyt, M. Tignor, and H. L. Miller, eds. "Contribution of Working Group I to the Fourth Assessment Report of the Intergovernmental Panel on Climate Change." Cambridge and New York: Cambridge University Press, 2007.

Speth, James Gustave. *Red Sky at Morning*. New Haven, CT: Yale University Press, 2004.

Spott, Pete. "Greenland's Ice Sheet: Climate Change Outlook Gets a Little More Dire." *Christian Science Monitor*, March 13, 2012.

Stensdal, Iselin. "China's Climate-Change Policy 1988–2011: From Zero to Hero?" Lysaker, Norway: Fridtjof Nansens Institutt, 2012.

Stone, Brian Jr. *The City and the Coming Climate: Climate Change in the Places We Live*. Cambridge: Cambridge University Press, 2012.

Stringer, C. B. "Evolution of Early Humans." In *The Cambridge Encyclopedia of Human Evolution*, edited by Steve Jones, Robert Martin, and David Pilbeam, 242. Cambridge: Cambridge University Press, 1994.

Subramanian, Meera. "The World's Fastest Animal Takes New York." Smithsonian.com, December 10, 2009. http://www.smithsonianmag.com/science-nature/The-Worlds-Fastest-Animal-Takes-New-York.html (accessed February 4, 2013).

Theil, Richard P. *Keepers of the Wolves: The Early Years of Wolf Recovery in Wisconsin*, Madison: University of Wisconsin Press, 2001.

Tir, Jaroslav, and Douglas M. Stinnett. "Weathering Climate Change: Can Institutions Mitigate International Water Conflict?" *Journal of Peace Research* 49 (2012): 211.

Tonnessen, J. N., and A. O. Johnsen. *The History of Modern Whaling*. London: C. Hurst and Company, 1982.

Truman, David B. "The State Delegations and the Structure of Party Voting in the United States House of Representatives." *The American Political Science Review* 50, no. 4 (1956): 1023–1045.

Tuchman, Barbara. *The March of Folly: From Troy to Vietnam*. New York: Random House, 1985.

Tyndall, John. *Contributions to Molecular Physics in the Domain of Radiant Heat*. New York: Appleton and Company, 1873.

Union of Concerned Scientists (UCS). "Life in the Slow Lane: Tracking Decades of Automaker Roadblocks to Fuel Economy." 2003. http://www.ucsusa.org/clean_vehicles/smart-transportation-solutions/better-fuel-efficiency/life-in-the-slow-lane.html (accessed March 21, 2014).

United Nations. "Climate Change 2014: Impacts, Adaptation and Vulnerability," IPCC, http://ipcc.ch/report/ar5/wg2/ (accessed July 30, 2014).

U.S. Congress. "Energy Independence and Security Act, 2007." Washington, D.C.: U.S. Government Printing Office, 2007.

U.S. Department of Energy (U.S. DoE). "Alternative Fuels Data Center." Washington, D.C., 2014. http://www.afdc.energy.gov/data/categories/vehicles (accessed July 28, 2014).

U.S. Department of State, Office of the Historian. "Milestones: 1977–1980: The Panama Canal and the Torrijos-Carter Treaties." https://history.state.gov/milestones/1977-1980/panama-canal (accessed March 19, 2014).

U.S. Department of State. *Protection of the Ozone Layer: Convention Between the United States of America and Other Governments*. Washington, D.C.: U.S. Government Printing Office, 1985.

U.S. Fish and Wildlife Service (USFWS). "Management of Wolf Conflicts and Depredating Wolves in Michigan," 2012. http://www.fws.gov/midwest/wolf/archives/depredation/mdnr_ea_qas.htm (accessed July 29, 2014).

U.S. Global Change Research Program (USGCRP). *Global Climate Change Impacts in the United States*, edited by Thomas R. Karl, Jerry M. Melillo, and Thomas C. Peterson. New York: Cambridge University Press, 2009.

U.S. House of Representatives, Committee on Science, Space, and Technology, Hearing Charter. *Climate Change: Examining the Processes Used to Create Science and Policy*. Thursday, March 31, 2011 10:00 a.m. to 12:00 p.m. 2318 Rayburn House Office Building.

U.S. National Park Service. "Glacier National Park." http://www.nps.gov/glac/naturescience/glaciers.htm (accessed March 11, 2013).

U.S. Senate. "Senate History: Treaties." http://www.senate.gov/artandhistory/history/common/briefing/Treaties.htm (accessed March 19, 2014).

Velicogna, Isabella, and John Wahr. "Acceleration of Greenland Ice Mass Loss in Spring 2004." *Nature* 443 (2006): 329–331. doi:10.1038/nature05168.

von Stein, Jana. "International Law and the Politics of Climate Change." *Journal of Conflict Resolution* 53, no. 2 (2008): 243–268.

Weart, Spencer. "Discovery of Global Warming." The American Institute of Physics (2013). http://www.aip.org/history/climate/floods.htm (accessed May 15, 2013).

Weiss, Edith Brown. Introduction: Vienna Convention for the Protection of the Ozone Layer, Vienna, March 22, 1985 and Montreal Protocol on Substances that Deplete the Ozone Layer, Montreal, September 16, 1987. Office of Legal Affairs, the United Nations, 2008. http: http://untreaty.un.org/cod/avl/ha/vcpol/vcpol.html (accessed December 7, 2012).

Whitesell, William C. *Climate Policy Foundations: Science and Economics with Lessons from Monetary Regulation*. New York: Cambridge University Press, 2011.

Wilson, David Sloan. *Darwin's Cathedral: Evolution, Religion, and the Nature of Society*. Chicago: University of Chicago Press, 2002.

World Bank. "New Report Examines Risks of 4 Degree Hotter World by End of Century." November 18, 2012. http://www.worldbank.org/en/news/press-release/2012/11/18/new-report-examines-risks-of-degree-hotter-world-by-end-of-century (accessed May 12, 2014).

World Wildlife Fund (WWF). "G8 Leaders Turn towards Copenhagen but Road Map Still Missing." 2009. http://wwf.panda.org/?169781/G8-climate-goal-Leaders-came-back-to-earth_But-what-now_-WWF (accessed July 28, 2014).

Wydeven, Adrian P, Randle L. Jurewicz, Timothy R. Van Deelen, John Erb, James H. Hammill, Dean E. Beyer Jr., Brian Roell, Jane E. Wiedenhoeft, and Daivd A. Weitz. "Gray Wolf Conservation in the Great Lakes Region of the United States." In *A New Era for Wolves and People: Wolf Recovery, Human Attitudes, and Policy*, edited by Marco Musiani, Luigi Boitani, and Paul C. Paquet. Calgary, B.C.: University of Calgary Press, 2009a.

Wydeven, Adrian P., Timothy R. Van Deelen, and Edward J. Heske, eds. *Recovery of Gray Wolves in the Great Lakes Region of the United States*. New York: Springer Press, 2009b.

Yaffee, Steven Lewis. *Prohibitive Policy: Implementing the Federal Endangered Species Act*. Cambridge: MIT Press, 1982.

INDEX

ABOUT THE AUTHOR

Patrick Regan is a professor of peace studies and political science at the University of Notre Dame. He has written three books and numerous articles on conflict and its management. His current research efforts focus on developing an understanding of the links between climate change and armed conflict. In addition to teaching at Notre Dame, he teaches a course in a medium security prison in Indiana.